Contemporary Research and Cutting-Edge Technology in Electrical Engineering

Contemporary Research and Cutting-Edge Technology in Electrical Engineering

Edited by
Dr. Suganyadevi M. V.
Dr. R. Ramya

CWP
Central West Publishing

A catalogue record for this book is available from the National Library of Australia

NATIONAL LIBRARY OF AUSTRALIA

ISBN (print): 978-1-922617-24-8

About the Editors

Dr. M. V. Suganyadevi obtained her Ph.D. from Anna University, Tamilnadu, India in 2015. Her research from 2009 to 2015 was based on voltage stability, where she developed the voltage stability margin using a simulation software. Specifically, she developed a new loadability margin index based on support vector regression and extreme learning machine for the estimation of voltage stability margin in online power grid. Currently, she is working as Associate Professor in Saranathan College of Engineering, Trichy, Tamilnadu, India. She has published 31 international journal articles and 29 conference papers. She is the member of IEEE, ISTE, ISRD and ISENG. She also acts as a reviewer for many SCI Journals. Her areas of research include power system stability, soft computing techniques and power quality.

R. Ramya obtained her Ph.D. in power systems in 2017 from Anna University Chennai, India. She is currently working as Assistant Professor in the Department of Electrical and Electronics Engineering in SRM Institute of Science and Technology, Chennai, Tamilnadu, India. Her research interests include machine modeling, power system optimization, real time simulation and power system control.

Contents

Preface

Preface

The objective of this book is to present cutting-edge technology in electrical engineering. The research areas as diverse as computer and communication networks; electronic circuits and systems; lasers and photonics; semiconductor and solid-state devices; nano-electronics; biomedical engineering; computational methods; artificial intelligence; robotics, design and manufacturing; control and optimization; computer algorithms; games and graphics; software engineering; computer architecture; cryptography and computer security; power and energy systems; and financial analysis; have been presented in sufficient depth to provide the students the basic theory as well as insights into applications. The approach is designed to develop the thinking process in the students, enabling them to reach a sound understanding of a broad range of topics related to electrical engineering, while motivating their interest in the electrical power industry.

1

A Roadmap to Smart Home Automation Sensors and Technologies

Sangeethapriya J.

Department of Information Technology, Saranathan College of Engineering, Tiruchirappalli, India

Abstract

Life is becoming simpler and easier in almost every way as automation technology advances. A wireless automation system that uses the internet to monitor simple home functionality and features from anywhere in the world is known as a smart home automation. Because of their various applications to users, smart home systems have gained importance. Smart homes, smart buildings, industrial Internet of Things (IoT) are those in which household appliances, electrical appliances, safety systems and computers can be monitored and controlled remotely. This chapter includes a roadmap on IoT, automation technology and various recommendations into smart home automation system sensors.

1.1 Introduction

The homes in the future will be completely automated and self-controlled because of its smart handling and other benefits. Industrial IoT automation system that enables the users to control the various kinds of electrical appliances. The wired communication is used by many existing domestic automation systems. This will not cause a problem until the system is planned and installed well in advance during the physical building. But the realization of cable networks costs for existing buildings in relation to other buildings is very high.

In the case of wireless systems, it can help for automation systems. Due to the advancement in technologies such as Wi-Fi, cloud networks in recent days, wireless systems are used everywhere than before. So, the wireless network is more advantageous and easily accessible than the wired network [11]. Sensors, displays, connectors, appliances, and devices that are connected together to enable domestic automation as well as local and remote control. Examples of controllable appliances and equipment are heating and heater systems (boilers, radiators), bright lights, windows, curtains, garage door, coolers, televisions and washers [1]. Smart automation technology must mature, and can be accomplished by defining, evaluating, and enforcing a broad variety of factors, both technical and non-technological. For example, the outline of wireless smart home systems is shown in Figure 1.1.

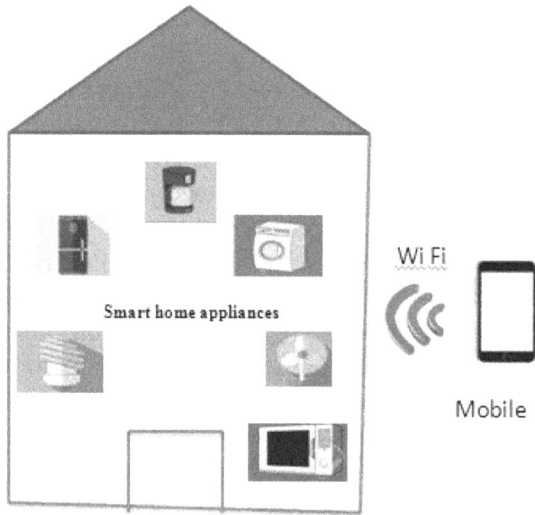

Figure 1.1 Wireless smart home system.

Benefits of Wireless Smart Automation System

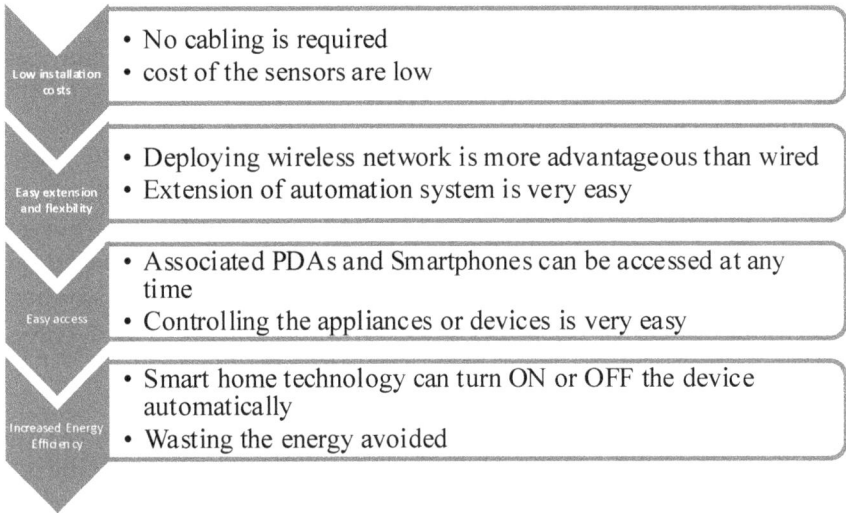

Low installation costs
- No cabling is required
- cost of the sensors are low

Easy extension and flexbility
- Deploying wireless network is more advantageous than wired
- Extension of automation system is very easy

Easy access
- Associated PDAs and Smartphones can be accessed at any time
- Controlling the appliances or devices is very easy

Increased Energy Efficiency
- Smart home technology can turn ON or OFF the device automatically
- Wasting the energy avoided

The advantages mentioned above, which keep the home secure with smart locks, Smart security warnings keep an eye on the building. With a smart thermostat, you can control the temperature in your home and save money on electricity. Control your home appliances such as the refrigerator, washing machine, air conditioner, lamps, fans, and microwave, among others [3]. Section 1.2 discusses the smart home system related background. Section 1.3 describes the various technologies and sensors adopted in the smart home automation system. The benefit and the risk of the smart home system are highlighted in Section 1.4. Section 1.5 presents the conclusion and acts as an eye opener to the researchers.

1.2 Related Works

The authors of [3] provided an overview of intelligent home systems, focusing on their architecture and general application areas, for energy conservation and home care. Furthermore, authors presented a brief overview of various home safety/security situations, but they did not go into detail about the mechanisms or authentication methods used. The authors of [12] used an IoT device identification and authentication learning algorithm and blockchain technology to take smart decisions. Smart decision-making and analytical skillful systems in need of replacement are desperately needed.

The authors of [7] discussed the design of the Internet of Things (IoT) and the architecture of Smart Homes that use IoT. In this paper, new methods and methodologies like CPLD controllers, zigbee modules, and RF modules have been discussed that are already being used to develop IoT applications. The paper [10] examined the various energy saving strategies used in the field of smart homes. Ordinary household appliances are used to incorporate this smart home system. With the support of the IFTTT (If This Then That) and Blynk applications, natural language voice commands are provided to the Google Assistant [5] in paper [7]. Three factors influenced the acceptability behavior of a smart home service: verification, connectivity and reliability. Customer behavior intentions have been tested by the researchers for smart home services. [2].

In order to meet the global requirements for automation and provide more convenient control and management, the smart Home is a home-based platform that connects various systems through the network. It is built on the Internet of things, information technology, control technology, imaging and communication technology.

1.3 Technologies and Sensors for Smart Automation System

Building a smart home entails a number of major components. The following are the main components: Smart Sensors, IoT Access points, IoT Protocols, IoT Hardware, IoT Fog computing and data warehouses are used to describe the Internet of Things. Several types of devices, such as a gas sensor, light dependent resistor (LDR), DHT, and Node MCU as a microcontroller, are used to implement smart home systems using IoT. Different technologies [6], including LoRaWAN, Server-based LoRa gateway, Bluetooth, Wi-Fi, RFID, zig-bee, cellular networks, (If-Then-That) IFTT and Blynk application [14], have been used to embed various levels of intelligence in the home, both short-range and long-range communications [1-3,8-10]. Table 1.1 shows six key element technologies used for different smart applications.

Table 1.1 Six key element technologies used for different smart applications

IoT Technologies for Smart Automation System	Type of Connections	Smart Applications
LPWAN (Low Power Wide area network)	Star	Industrial IoT production and control, Smart Meter, Smart City, Smart Construction, Fitness trackers, Logistics & Asset Tracking, Smart Agriculture
Cellular	Star	Industrial IoT, Fitness, Smart Rental, Logistics & Asset Tracking.
Zigbee	Mostly Mesh	Industrial IoT, Smart Building, Smart Home
BLE (Bluetooth Low Energy)	Star & Mesh	Smart Home, Smart Building, Wearable's, Smart Health, Smart Rental.
Wi-fi (Wireless Fidelity)	Star & Mesh	Smart Home, Connected Car, Smart Rental
RFID (Radio Frequency Identification)	Point to Point	Smart Agriculture, Logistics & Asset Tracking.

IoT Sensors [2] are playing vital role in Smart home technology. Based on their sensing capabilities the following sensors are very useful to build smart home automation systems. Table 1.2 depicts the different sensors and their applications [12].

Table 1.2 Different sensors and their applications [12].

IoT Sensors	Applications
Pressure sensors	Ammonia leakage from refrigerator, air flow in air duct filter
Temperature sensors	Solar heating pump, Air control conditioning
LPG sensors	Sensing Concentration of LPG in air
Humidity sensors	Measure air moisture and used to switch on AC
Water level sensors	To check overflow of water in water

	tank and also in RO
Video camera for Surveillance	To monitor kith and kin
Motion sensors	To capture the footage of intrusion
Infrared sensors	To detect body heat, home security, microwave oven

1.4 Benefits and Risks

Environmental variables such as temperature, light, motion, and humidity are detected by sensors and monitors. Apps like Alexa [13] on processing devices (smart phones, tablets, laptops, Personal digital assistant) [4] or dedicated firmware devices (e.g., wall-mounted controls) provide control features. Home automation has a number of incredible (and undeniable) benefits [1]. They are:

- Makes easier to access and managing all appliances
- Smart home technology is easily upgraded by latest technology
- Raising the level of protection in your home.
- Home appliance features can be controlled remotely.
- Energy performance is improved.
- Home appliances performance has been improved.

Besides, smart-home systems are vulnerable to a variety of security flaws [1] that, if not addressed, can put your information or services at risk.

- Smart-home devices contain a wealth of personal information that cyber criminals can steal by hacking if they lack solid protection against attacks from your birth date to credit card details.
- A Weak password allows smart hackers to easily access and manipulates the hub and other smart devices at home.
- Breaches of intelligent devices, such as cooling and heating that monitor critical functions at home can be even worse.
- Upgradation of software is not properly then prone to viruses.
- Dependency on latest technologies
- Dependency on electricity power
- Makes the people more lazy

1.5 Conclusion

Intelligent homes are a focus field of national and strategic energy planning. The adoption on the market of intelligent home technologies is based on future users who experience clear advantages with acceptable risk levels. If the company provides adequate security control and analysis in the IoT market, safe communication, good mutual authentication, secure boot and lifecycle management at an affordable price make our intelligent home technology safer and more popular. The increasing burden on stakeholders is currently important in terms of energy consumption. Researchers can focus on energy constraint mechanisms, security techniques and develop new sensors for newer applications that use secure protocols.

References

1. Wilson, C., Hargreaves, T., and Hauxwell-Baldwin, R. (2017) Benefits and risks of smart home technologies. *Energy Policy*, **103**, 72-83.
2. Yang, H., Lee, W., and Lee, H. (2018) IoT smart home adoption: The importance of proper level automation. *Journal of Sensors*, 11-12.
3. Fadhil, J. A., Omar, O. A., and Sarhan, Q. I. (2020) A Survey on the Applications of Smart Home Systems. *International Conference on Computer Science and Software Engineering (CSASE)*, 168-174.
4. Mandula, K., Parupalli, R., Murty, C. H. A. S., Magesh, E., and Lunagariya, R. (2015) Mobile Based Home Automation using Internet of Things (IoT). *International Conference on Control, Instrumentation, Communication and Computational Technologies (ICCICCT)*, 340-343.
5. Gupta, M. P. (2018) Google Assistant controlled home automation. *International Research Journal of Engineering and Technology (IRJET)*, **05**, 2074-2078.
6. Li, M., Gu, W., Chen, W., He, Y., Wu, Y., and Zhang, Y. (2018) Smart home: Architecture, technologies and systems. *Procedia Computer Science*, **131**, 393-400.
7. Gaikwad, P. P., Gabhane, J. P., and Golait, S. S. (2015) A Survey Based on Smart Homes System using Internet-Of-Things. *International Conference on Computation of Power, Energy, Information and Communication*, 330-335.
8. Islam, R., Rahman, Md. W., Rubaiat, R., Hasan, Md. M., Reza, Md. M., and Rahman. Md. M. (2021) LoRa and server-based home automation using the internet of things (IoT). *Journal of King Saud Univer-*

sity   -   Computer   and   Information   Sciences, https://doi.org/10.1016/j.jksuci.2020.12.020.

9. Majeed, R., Abdullah, N. A., Ashraf, I., Zikria, Y. B., Mushtaq. Md. F., and Umer, Md. (2020) An Intelligent, Secure, and Smart Home Automation System. *Scientific Programming Towards a Smart World 2020*, https://doi.org/10.1155/2020/4579291.

10. Suresh, S., and Sruthi, P. V. (2015) A Review on Smart Home Technology. *International Conference on Green Engineering and Technologies*, https:// 10.1109/GET.2015.7453832.

11. Vikram, N., Harish, K. S., Nihaal, M. S., Umesh, R., and Kumar. S. A. A. (2017) A Low Cost Home Automation System Using Wi-Fi Based Wireless Sensor Network Incorporating Internet of Things (IoT). *IEEE 7th International Advance Computing Conference (IACC)*, https://doi.org/10.1109/IACC.2017.0048.

12. https://dzone.com/articles/home-automation-using-iot

13. https://iotcircuithub.com/alexa-home-automation-esp8266

14. https://how2electronics.com/iot-home-automation-using-blynk-nodemcu-esp8266/

2

Development of Wind and Solar Based AC Microgrid With Power Quality Improvement for Local Nonlinear Load

M. V. Suganyadevi[1] and T. Thenmozhi[2]

[1]Department of EEE, Saranathan College of Engineering, Tamilnadu, India
[2]Department of EEE, Sri Sai Ram Institute of Technology, Tamilnadu, India

Abstract

This project proposes a microgrid (μ-grid) which is been integrated with the wind also as solar photovoltaic (PV) resources, alongside the battery energy storage (BES) to the three-phase grid which is feeding to the nonlinear load. The μ-grid disconcerted by probabilistic nonlinear time dependent parameters and their effects are compensated by cohesive controllers used for utility grid side voltage source converter (GVSC) and machine side voltage source converter (MVSC). The nonlinear load compensation and therefore the PQ enhancement are achieved by executing a modified version which is an adaptive filtering technique called "momentum"-based least mean square (MLMS) control technique, which is especially utilized for providing the switching control signals to the GVSC. The MVSC acquires its switching signals from conventional vector control scheme and therefore the encoder less estimation of speed and rotor position of the synchronous generator driven by turbine through back electromotive force control technique.

2.1 Introduction

With rapid developments within the industry, power quality becomes vital. Power quality is defined as any power problem manifested within the voltage, current or frequency deviations that end in failure or mal function of the customer equipment. Power quality issues are often classified as short duration voltage variations, long duration voltage variations, waveform distortions, transients, voltage imbalance and voltage flicker. Among the varied power quality issues, voltage sag, voltage swell and harmonics are more dominant within the distribution system. The expansion of economies has led to the rise of energy demand. By 2040, the demand may double or even triple as an outcome of population rise. The conservation of energy, research on renewable energy resource (RERs) applicability and halting the dependency on fossil fuels, is of utmost importance. The RERs are dominated by their intermittency and geographic location availability. The cost of the wind and solar power generation has been rapidly falling since the last decade. Driven by their economic and technical incentives, the worldwide installed capacity of photovoltaic (PV) and wind generators has approached 303 Gigawatt (GW) and 487 GW in 2016, as compared to 6 GW and 74 GW in 2008, respectively.

A standalone wind-PV cogeneration system is proposed. On the small-scale level, a single-phase cogeneration system has been proposed whereas a laboratory-scale system is introduced. Generally, the system structure comprises a common dc-bus that interfaces several parallel connected converters-interfaced renewable energy resources, which might reduce the overall system efficiency and increase the cost. More importantly, the cascaded connection of power converters requires rigorous controller's coordination to avoid the induced interactions dynamics, which could yield instabilities. A back-to- back (BtB) voltage-source converter (VSC) connected to a doubly fed induction generator is employed to interface a dc-dc converter-interfaced PV generator and an energy storage unit. A PV generator charging A battery bank and interfaced to a wind driven induction generator via a VSC is proposed. The wind-PV cogeneration systems highlight the efficient integration of the renewable energy resources with the minimal utilization of power-electronic conversion stages... A buck/buck-boost fused dc-dc converter is proposed. A dc-dc converter with a current-source interface, and a coupled transformer is proposed respectively. The system comprises a BtB VSCs to interface the PV and wind generators to the utility-grid. On the machine-side- VSC, the dc-link voltage is regulated to the utmost point tracking (MPPT) value of the PV panels by an outer loop proportional-and-integral (PI) dc voltage controller. The reference values of the machine-side currents are calculated using the synchronous detection method, and a hysteresis current controller is utilized for the regulation. On the grid-side-VSC, a hysteresis grid-current controller is used to inject the total currents into the utility-grid. In spite of the potential benefits of the proposed system.

2.2 System Architecture and Control

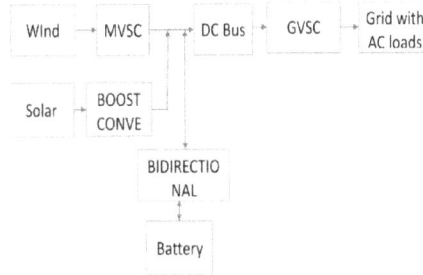

```
Wind ──► MVSC ──┬─►  DC Bus ──► GVSC ──►  Grid with
                │                          AC loads

       Solar ─► BOOST
               CONVE │
                     │
                BIDIRECTIO
                   NAL
                     │
                  Battery
```

The wind and solar based AC microgrid system where two back-to-back associated VSCs viz. MVSC and GVSC provide the decoupled control of the electrical and the mechanical power.

A solar PV array (single stage) is associated to the DC link through the BES and bidirectional DC-DC converter. The load demand is met by the combined generation from wind turbine driven SG and solar PV array. The nonlinear load is connected at point of common interconnection (PCI). The grid engrosses the excess generation from solar PV array and wind generation unit. The switching harmonics, i.e., the power quality, are filtered using a ripple filter and control techniques connected at PCI. The integration of the power from renewables, i.e., wind and solar into the utility grid without hampering the reliability of the supply is the basic objective. The BES allows the favorable solution of mitigation of the fluctuations caused by the generation from renewables. The switching of VSCs with the proven control algorithms lead to the improvement in the power quality as stated by the IEEE-519 standard.

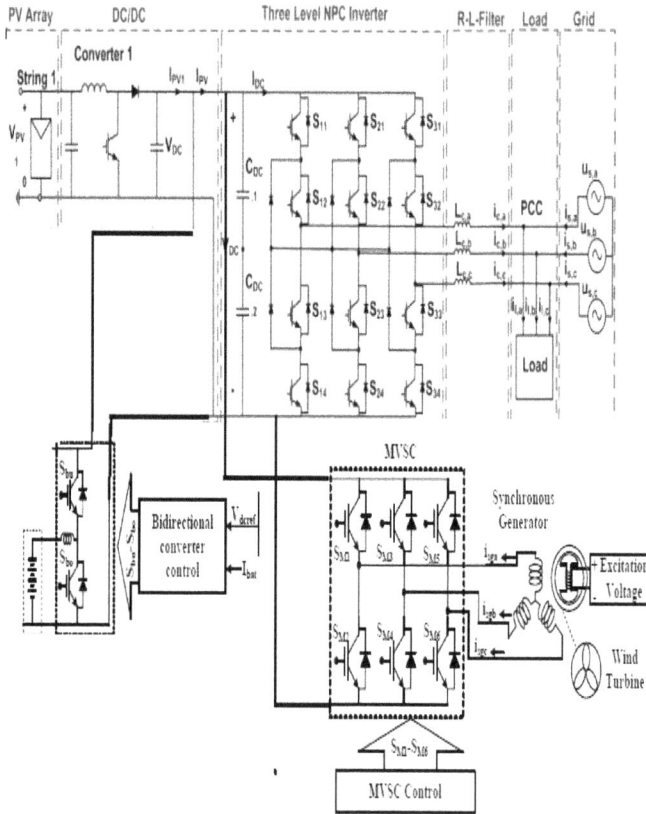

Figure 2.1 Overall circuit diagram.

The circuit diagram consists of VSR and VSI converter. The VSR converter is controllable rectifier which converts the wind AC power to DC power which will be connected to the DC link. The solar power is connected to the DC link via DC-DC boost converter in which the duty cycle is generated by the INC based MPPT controller. The bidirectional converter fed by the battery source which increases the reliability of the system during nighttime when solar is absent. The DC link has the capability of delivering to the AC load as well as grid. The VSI inverter takes the role of converting DC to AC power which is also controllable inverter which increases the flexibility of AC transmission. The switching pulses to the VSI is given by the d-q based control which increases the stability of the voltage.

A. VSR Control System

The proposed VSR control system includes the VDC regulator, PLL measurement unit as well reference voltage generation. The VDC regulator has the feedback of dc link voltage measured value. The measured value is compared with the reference value and the PI controller is used to generate the reference value of current in d-axis i.e., Id. The PLL measurement block is used to convert the Vabc and Iabc of wind to the d-q component. The PLL block is also used to find the ωt value for converting the abc to dq value.

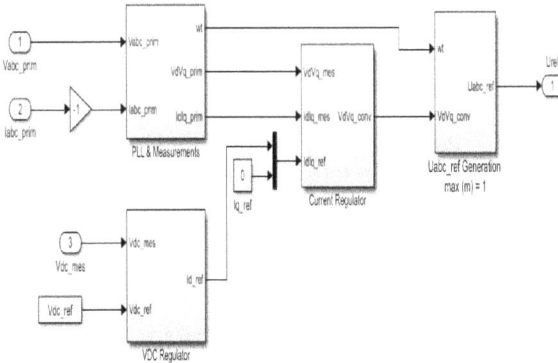

Figure 2.2 VSR control strategy.

The converted dq value and Id ref are given to the current regulator which generates the Vd&Vq reference value. The reference value of d-q axis voltage component is fed to the reference abc voltage generation block. The Uref is given to the PWM generator, which generates required switching signal to the VSC1. The PWM generator will compare the reference voltage signal with the carrier signal with specified frequency. The P&O based MPPT is utilized for maximum power extraction from wind turbine. The wind MPPT generates the reference speed of the generator.

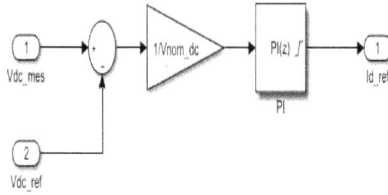

Figure 2.3 VDC control.

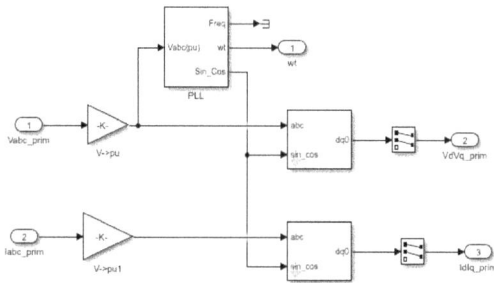

Figure 2.4 PLL measurement and D-Q conversion.

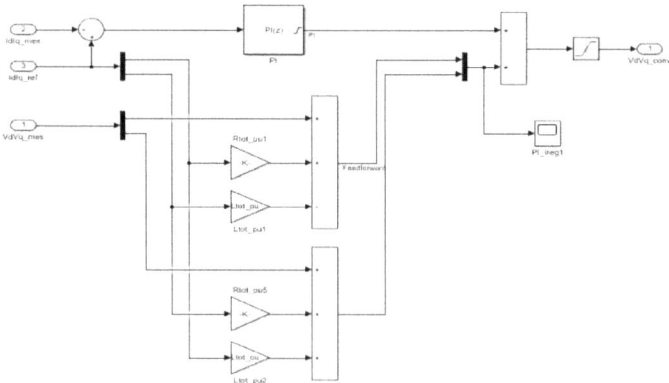

Figure 2.5 Current regulator for reference Vd &Vq generation.

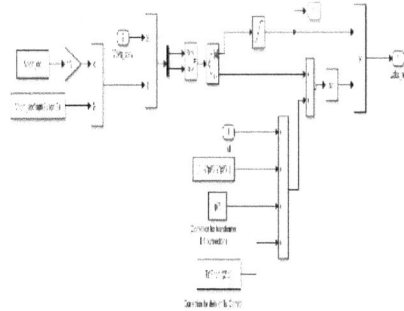

Figure 2.6 Reference Uabc generation.

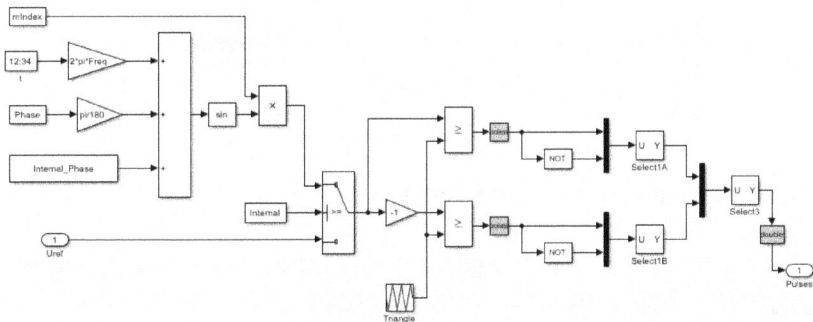

Figure 2.7 Pulse generator.

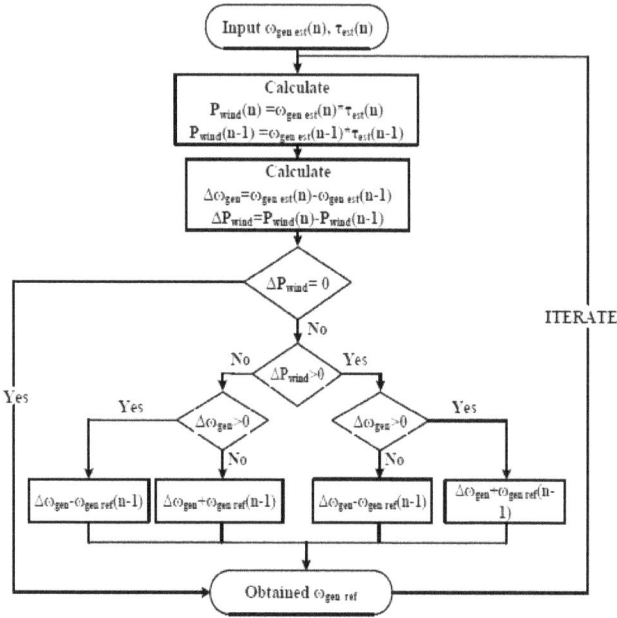

Figure 2.8 Flowchart of P&O based MPPT for wind.

The control deals with the generation of the reference direct axis (Id ref) and quadrature axis component (Iq ref) of the generator current. The quadrature axis component is the replica of the torque producing component. The wind-P&O MPP scheme provides the reference generator speed (ωgen ref) and is compared with the estimated SG speed (ωgen est) obtained by the encoderless BEMF technique.

B. VSI Control Strategy

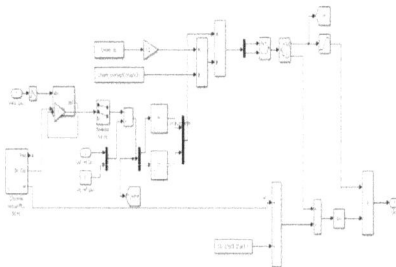

Figure 2.9 Control scheme for VSI.

The load voltage is taken as input, and they are converter D-Q component by parks transformation. Figure 2.9 shows the control scheme for VSI.

Park Transformation

The abc_to_dq0 transformation block computes the direct axis, quadratic axis, and zero sequence quantities in a two-axis rotating reference frame for a three-phase sinusoidal signal. The following transformation is used:

$$V_d = \frac{2}{3}\left(V_a \sin(\omega t) + V_b \sin(\omega t - 2\pi/3) + V_c \sin(\omega t + 2\pi/3)\right)$$

$$(1)$$

$$V_q = \frac{2}{3}\left(V_a \cos(\omega t) + V_b \cos(\omega t - 2\pi/3) + V_c \cos(\omega t + 2\pi/3)\right)$$

$$(2)$$

$$V_0 = \frac{1}{3}\left(V_a + V_b + V_c\right)$$

$$(3)$$

where ω = rotation speed (rad/s) of the rotating frame.

The transformation is the same for the case of a three-phase current; you simply replace the Va, Vb, Vc, Vd, Vq, and V0 variables with the Ia, Ib, Ic, Id, Iq, and I0 variables. This transformation is commonly used in three-phase electric machine models, where it is known as a Park transformation. It allows you to eliminate time-varying inductances by referring the stator and rotor quantities to a fixed or rotating reference frame. In the case of a synchronous machine, the stator quantities are referred to the rotor. Id and Iq represent the two DC currents flowing in the two equivalent rotor windings (d winding directly on the same axis as the field winding, and q winding on the quadratic axis), producing the same flux as the stator Ia, Ib, and Ic currents.

One can use this block in a control system to measure the positive-sequence component V1 of a set of three-phase voltages or currents. The Vd and Vq (or Id and Iq) then represent the rectangular coordinates of the positive-sequence component.

The converted d-q component is compared with the reference value of d-q component through the PI controller.

PI Controller

A variation of Proportional Integral Derivative (PID) control is to use only the proportional and integral terms as PI control. The PI controller is the most popular variation, even more than full PID controllers. The value of the controller output is fed into the system as the manipulated variable input.

$$e(t) = SP - PV \tag{4}$$

$$u(t) = u_{bias} + K_c\, e(t) + \frac{K_c}{T_I} \int_0^t e(t)dt \tag{5}$$

The u_{bias} term is a constant that is typically set to the value of u(t) when the controller is first switched from manual to automatic mode. This gives "bump less" transfer if the error is zero when the controller is turned on. The two tuning values for a PI controller are the controller gain, K_c and the integral time constant, T_I. The value K_c is a multiplier on the proportional error and integral term and a higher value makes the controller more aggressive at responding to errors away from the set point. The set point (SP) is the target value and process variable (PV) is the measured value that may deviate from the desired value. The error from the set point is the difference between the SP and PV and is defined as e(t)=SP-PV.

Digital controllers are implemented with discrete sampling periods and a discrete form of the PI equation is needed to approximate the integral of the error. This modification replaces the continuous form of the integral with a summation of the error and uses Δt as the time between sampling instances and n t as the number of sampling instances.

$$u(t) = u_{bias} + K_c\, e(t) + \frac{K_c}{T_I} \sum_{i=1}^{n_t} e_i(t)\Delta t \tag{6}$$

PI control is needed for non-integrating processes, meaning any process that eventually returns to the same output given the same set of inputs and disturbances. A P-only controller is best suited to integrating processes. Integral action is used to remove offset and can be thought of as an adjustable u_{bias}..

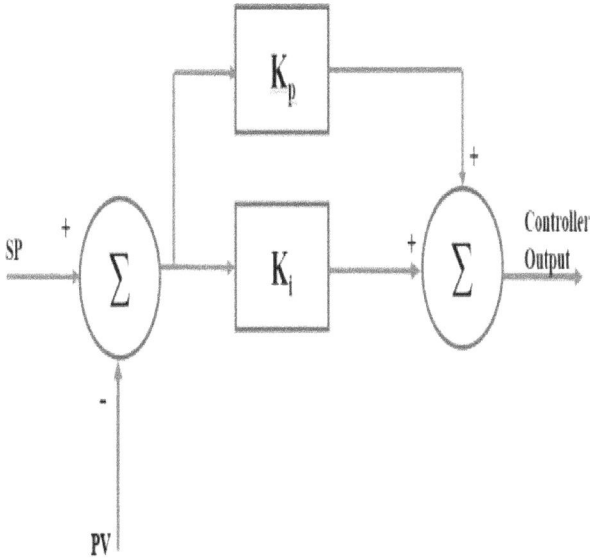

Figure 2.10 PI block diagram.

The control voltage signal is divided with the nominal voltage. After, it is converter to real and imaginary values. The phase angle and magnitude are divided separately. The phase angle is corrected as per the fundamental frequency 50 Hz and the magnitude of d-q component is again converted into abc component which is given as reference modulated signal for the reference level generation.

The P&O based MPPT is used for PV maximum power extraction which generated the nominal voltage value which is given to the VSI control strategy as Vnom_dc feedback value. The adaptive P&O MPP control overcomes the oscillations issues of fixed step based MPP techniques by introducing a variable perturbation step size. The control estimates the MPP operating point by using a product of short circuit current and optimal proportionality constant. The tuning procedure of the perturbation step size is based on irradiance level and operating point oscillations around the power maxima named as coarse and fine tuning, respectively. For coarse tuning, the insolation level of solar PV array decides the perturbation size, whereas, in fine tuning, the oscillations around the MPP determines the perturbation size.

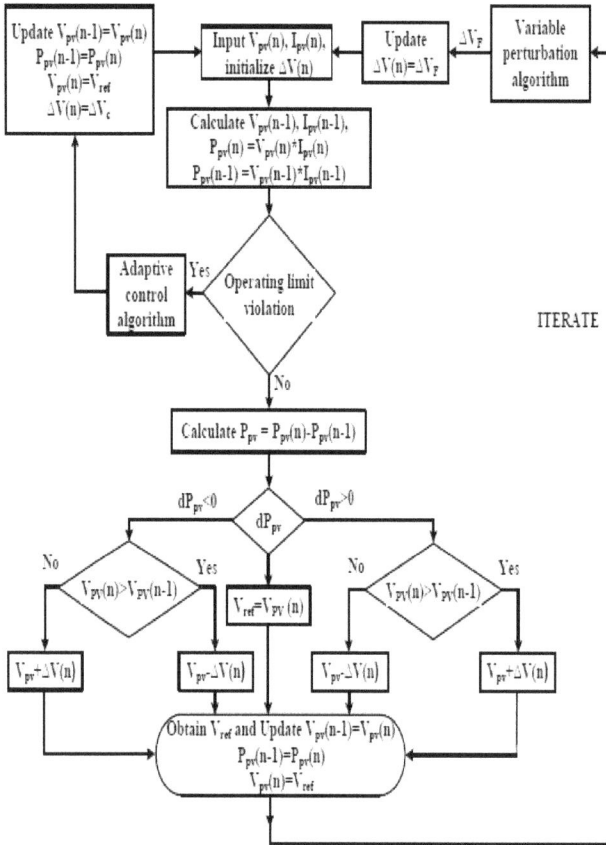

Figure 2.11 Flowchart of P&O based MPPT for PV.

2.3 Neutral Point Clamped Inverter

A neutral-point clamped (NPC) inverter is a three-level voltage-source inverter. The NPC topology has been adopted for high power applications as it can achieve better harmonic reduction than traditional two-level voltage source inverters and the associated control strategies help to minimize semiconductor losses. NPC type multi-level inverters plays vital role in the field of power electronics and being extensively used in various industrial and commercial applications because it possesses low electromagnetic interference, and the efficiency is considerably high. NPC Multilevel inverters have become more favored over the years in electric high-power application with the affirmation of less disturbances and the contingency to operate at lower switching frequencies than typical two-level inverters. This multilevel inverter will also be compared with two-level inverter in simulations to investigate the advantages of using multilevel inverters. It is observed that NPC multilevel inverter produce only 22% and 32% voltage THD whereas the two-level inverter for the same test produces 115% voltage THD. The typical voltage source inverters give the output voltage at the poles with levels +Vdc/2 or –Vdc/2 which is the DC link voltage also known as two level inverters. To attain a quality output voltage or current waveform with minimal number of ripple content, it needs high switching frequency along with other discrete pulse width modulation techniques. In high voltage and power application the two-level inverters have some constraint in operating at high frequency mainly because of switching losses and restriction of equipment rating.

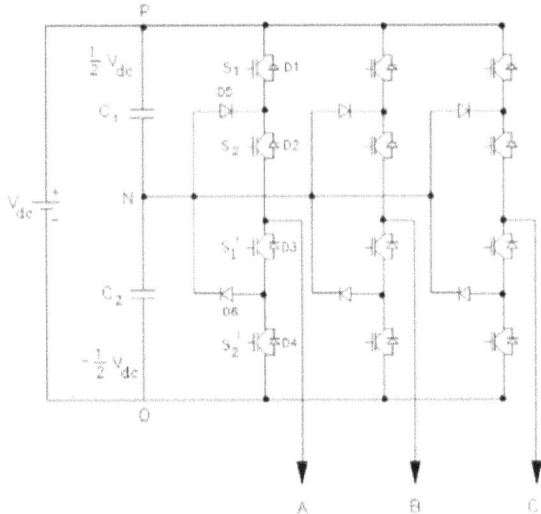

Figure 2.12 NPC inverter.

The three-level NPC inverter, widely used in applications for a three-phase three-wire system, originally has the structure of split dc capacitors. The existing dc neutral point can be directly utilized as the ground return. Actually, the three-level NPC inverter can be used in applications for a three-phase three-wire system and for a three-phase four-wire system.

2.4 Simulation Results

Overall simulation diagram

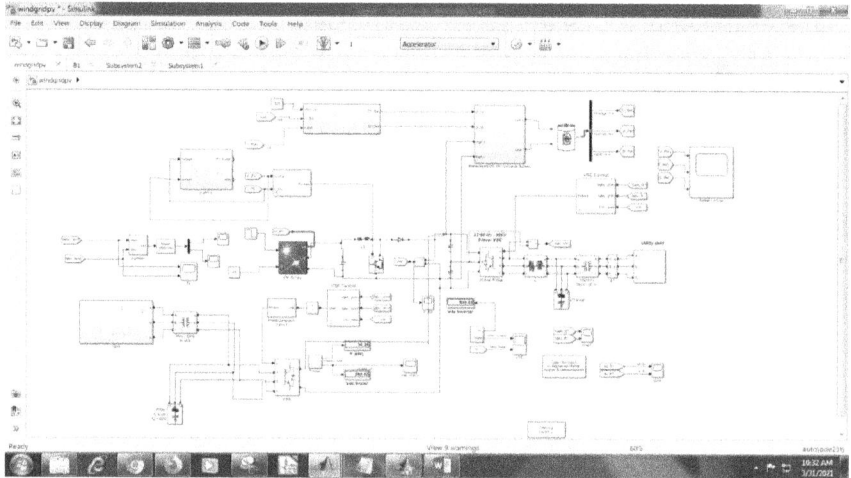

PV waveforms 1000 irradiance for 0-0.5 sec and 400 irradiance for 0.5-1.0 sec

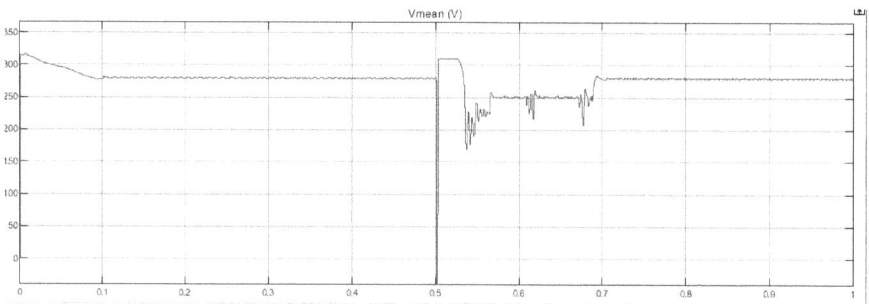

Current waveform of PV

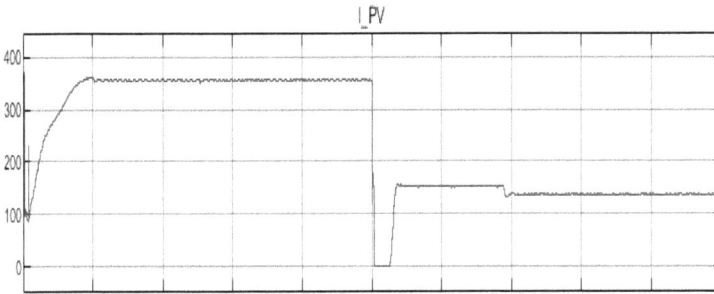

Power waveform of PV

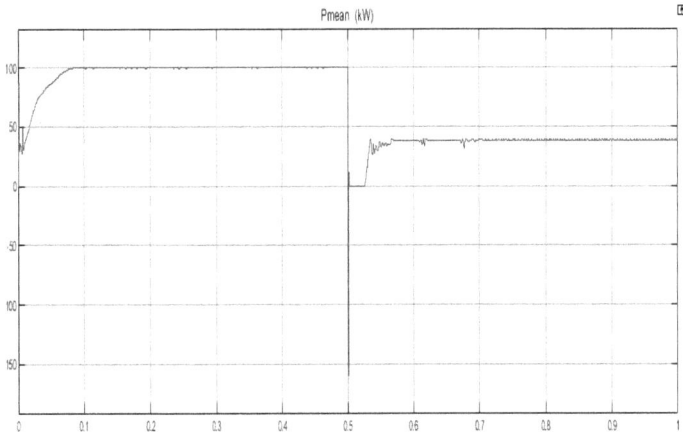

Battery waveforms during power variations of the wind and solar

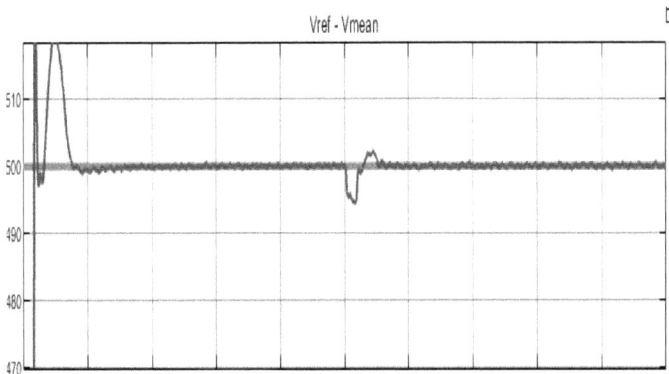

Common dc link voltage

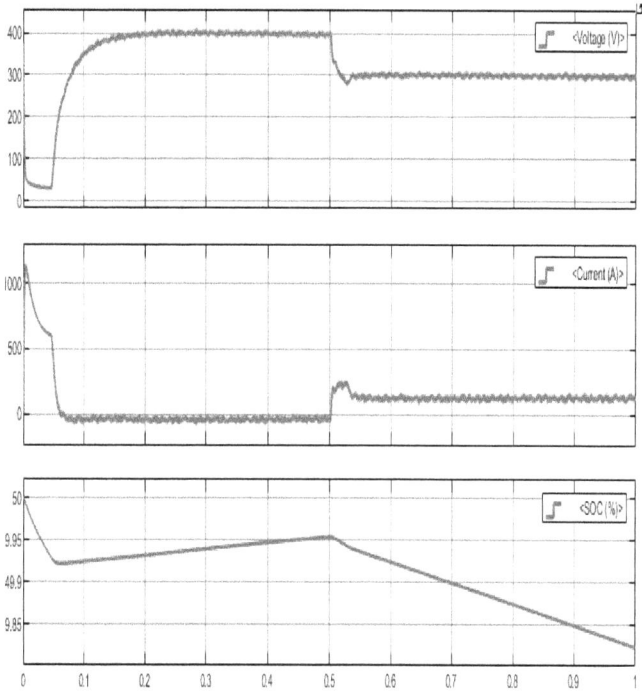

Wind voltage and current

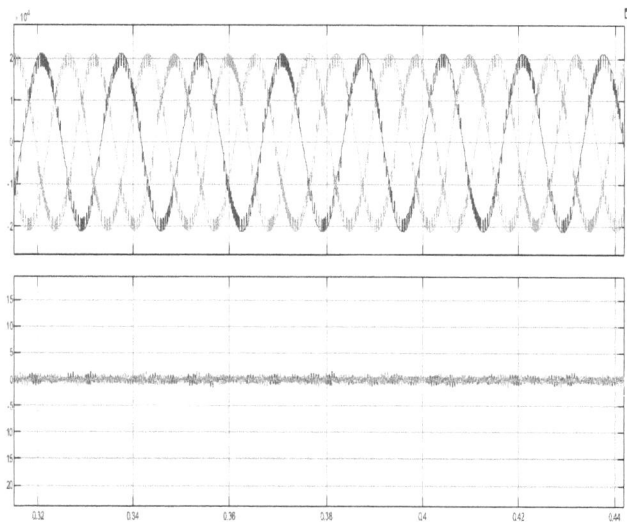

Grid voltage and current

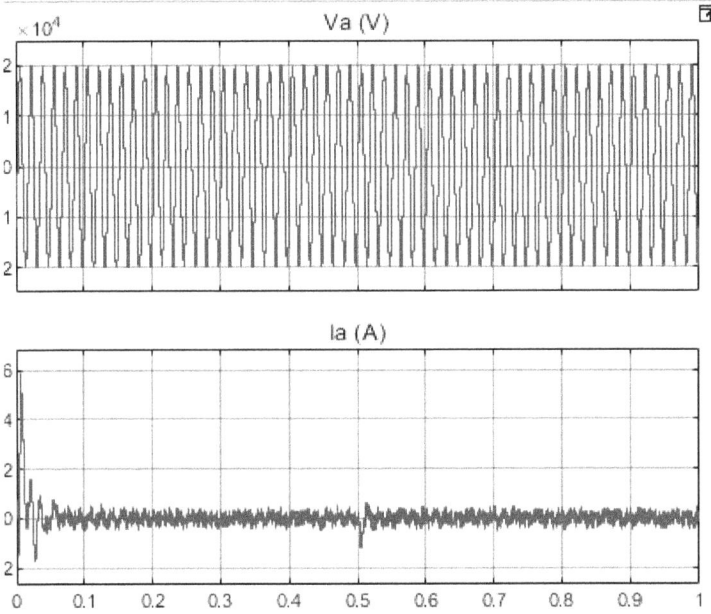

2.4 Conclusion

The proposed project implements hardware prototype the proposed wind-solar AC micro grid has been designed and implemented to illustrate its improved PQ performance for local nonlinear load. The main advantages of the proposed system are simple and efficient adaptive control and fast response. The proposed wind-power co generation with back-to-back converter VSC1 and VSC2 is implemented with two control strategy for converters. The wind-solar hybrid power resource increases the stability of power distribution of the system. The control strategy for two controllers will reduce the DC-DC conversion staged as well as accurate regulation of DC link voltage. The control techniques are simple structure and easier. The power conversion using back-to-back converter with proposed control strategies will have less harmonic distortions which have been verified using FFT analysis in Matlab/Simulink. The load/grid current THD has been found well within the IEEE-519 harmonic standard. The proposed system has operated well under all the dynamic conditions as well as the power quality issues are mitigated satisfactorily.

References

1. Blaabjerg, F., Chen, Z., and Kjaer, S. B. (2004) Power electronics as efficient interface in dispersed power generation systems. *IEEE Trans. Power Electron.*, **19**(5), 1184-1194.
2. Carrasco, J., et al. (2006) Power-electronic systems for the grid integration of renewable energy sources-a survey. *IEEE Trans. Ind. Electron*, **53**(4),1002-1016.
3. Nousiainen, L., Puukko, J., Maki, A., Messo, T., Huusari, J., Jokipii, J., Viinamaki, J., Lobera, D., Valkealahti, S., and Suntio, T. (2013) Photovoltaic generator as an input source for power electronic converters. *IEEE Trans. Power Electron.*, **28**(6) 3028-3038.
4. Hemapriya, C. K., Suganyadevi, M. V., and Krishnakumar, C. (2020) Detection and classification of multi-complex power quality events in a smart grid using Hilbert–Huang transform and support vector machine. *Electr. Eng.*, **102,** 1681-1706.
5. Strachan, N., and Jovcic, D. (2010) Stability of a variable-speed permanent magnet wind generator with weak ac grids. *IEEE Trans. Power Delivery*, **25**(4) 2279-2788.
6. Nagarajan, A., Sivachandran, P., Suganyadevi, M. V., and Muthukumar, P. (2020) A study of UPQC: emerging mitigation techniques for the impact of recent power quality issues. *Circuit World*, **47**(1), 11-21.
7. Mitra, P., Zhang, L., and Harnefors, L. (2014) Offshore wind integration to a weak grid by VSC-HVDC links using power-synchronization control - a case study. *IEEE Trans. Power Del.*, **29**(1), 453-461.
8. Wang, Y., Meng, J., Zhang, X., and Xu, L. (2015) Control of PMSG-based wind turbines for system inertial response and power oscillation damping. *IEEE Trans. Sustain. Energy*, **6**(2)565-574.
9. Suganyadevi, M. V., and Thenmozhi, T. (2021) Mitigation of sag/swell in transmission by using DVR. *Journal of Next Generation Technology*, **1**(1), 8-14.
10. Xu, L., Ruan, X., Mao, C., Zhang, B., and Luo, Y. (2013) An improved optimal sizing method for wind-solar-battery hybrid power system. *IEEE Trans. Sustain. Energy*, **4**(3) 774-785.
11. Sarkar, S., and Ajjarapu, V. (2011) MW resource assessment model for a hybrid energy conversion system with wind and solar resources. *IEEE Trans. Sustain. Energy*, **2**(4) 383-391.

3

Digital Controller for DC-DC Power Converter

S. Vijayalakshmi[1], R. Shenbagalakshmi[2], M. Marimuthu[3] and R. Venugopal[4]

[1,3,4]*Department of Electrical & Electronics Engineering, Saranathan College of Engineering, Trichy, Tamilnadu, India*
[2]*Department of Electrical & Electronics Engineering, Sinhad Science of Technology, Lonavala, Pune, Maharashtra, India*

Abstract

This chapter explains about a DC to DC converter which proceeds a DC source voltage and generates a different degree of output DC voltage by changing the duty ratio of the converter. The DC to DC power converter is similar to a transformer will increase or decrease a DC voltage of input. Buck and boost converters perform a significant role in DC to DC converters; other converters like buck-boost, Zeta, SEPIC, and so on might be obtained from these converters. These converters are commonly used in PV cells, electric cars, batteries, and automotive applications, among other things. It has a high performance, a smooth acceleration regulation, and a fast dynamic response. Particular key drawbacks of the DC/DC converters are the management of output voltage; hence several control techniques have been developed to address this limitation.

The control technique necessitates precise simulation and a detailed review of the converters. The state space average modelling is used for modelling the DC to DC converters. It uses the state-space definition of dynamical setup to extract the derivations of low signal average of the switching of PWM converters and allows simulation and control design easier. In general, DC/DC switching power converters are nonlinear and invariant in time and can be controlled using either linear or nonlinear control. The Proportional-Integral-Differential (PID) controller is a linear controller that can provide sufficient control action for DC-DC converters by varying the three constraints in the PID algorithm of controller. The output specification of the controllers can be defined with regard to the controller's sensitivity to an error, the step to that the regulator peak overshoots at the fixed point, and the step of system oscillation. PID controllers have the best control dynamics, with the least steady-state error, the fastest reaction (rise and settling time), no oscillations, and better stability.

It does not function reliably and adequately in the event of non-linear systems, it has a longer rising time when output voltage overshoot diminishes, and it suffers from dynamic response and causes overshoot, impairing converter output voltage regulation. The digital controller solves the problems described above in analogue controllers and has several advantages over their analogue counterparts. Some of its benefits include: i) digital materials are less vulnerable to ageing and environmental variations. ii) less prone to noise iii) has better hardware compatibility, and therefore the same configuration is adaptable to some kind of adjustment in the controller iv) versatility in changing controller characteristics v) ease of operation. It also offers greater flexibility, quicker response, and less overshoot in its dynamic reply.

The aim of this chapter is to develop a robust digital compensator for a DC/DC converter based on a discrete PID controller. The architecture is performed in the time domain, and the converter parameters are observed and evaluated. The open loop transfer feature of the constructed DC to DC converter circuit is found by modelling them by means of the state-space averaging technique. To develop a digital controller, an analogue PID controller is first designed for a DC-DC converter and then converted into a controller of discrete PID, and the Ziegler Nichols procedure is used to calculate the analogue PID controller constants for the DC-DC converter, which are K_P, T_D, and T_I.

In this segment, the analogue PID controller for a DC/DC converter is converted into a discrete compensator using a bilinear transformation technique, which removes errors associated with the conversion from continuous to discrete. The discrete feedback control loop varies the compensator factors constantly and simultaneously to attain the converter requirements. The suggested controller gives full compensation, which increases the robustness and reliability of the closed loop DC-DC converter even further. Among the various strategies available for transforming analogue controllers to digital controllers, bilinear transformation is chosen for conversion because it is more suitable for constructing all types of filters and compensators.

Furthermore, assured reliability is a desirable performance for any controller style.

The three areas of interest in the modern approach found in the study are i) Conversion of Analog to Digital (ADC) is analyzed to sample the voltage error, ii) A discrete compensator is used to compensate for the error signal. iii) The DPWM (Digital Pulse Width Modulation) signals, which must be produced with the uppermost resolution in order to achieve maximum precision in the appropriate output voltage. High resolution DPWM signals are provided in this proposed controller to maintain the maximum switching frequency. To control the controller's target is to run the DC to DC converter switches to a duty ratio such that the portion of the dc voltage output equals the reference voltage. Irrespective of variations in the voltage, which is given in the input or load, the control should remain unchanged.

3.1 Introduction

Figure 3.1 depicts an overview of the controller. The aim of this chapter is to create Discrete PID controllers for various DC-DC converters. Every DC to DC converter must be constructed and can be modelled using the averaging state-space technique. The converter transfer function could be derived from the modelling phase. Analog PID controller has been designed for the derived transfer function using Ziegler-Nichols tuning method.

The DC to DC converters are non-linear and variation in time of the dynamic devices that transform one degree of DC voltage to another by changing the converter's duty ratio. This kind of converters are more compact and have an efficiency range of 76% to 97%. The benefits of DC-DC converters include higher performance, flexibility in nature, less tension on the power switch, and only need a minor output filter for removing little output ripple.

In the world of power electronics, control elements are the most difficult to incorporate in DC-DC converters. The control method necessitates accurate emulation and a thorough examination of the converters. PID controller systems are suitable for controlling DC-DC converters. In the architecture, there are analogue PID controllers and optical PID controllers. The following issues must be resolved in traditional analogue PID controller approaches: poor stability, high rigidity to adjust advanced functions and device changes, and complexity in performance control. The optical PID controller architecture has many benefits over its analogue equivalent. Any of its benefits include: i) discrete components are less vulnerable to ageing so that there are some changes in the environment ii) little prone to noise iii) has greater hardware compatibility, so the same configuration can be adapted to suit any improvements in the controller iv) increased sensitivity to constraint changes v) the adaptability in the controller features vi) simplicity of action. In its dynamic response, a digital PID controller offers reliability, quicker response, and reduced overshoot.

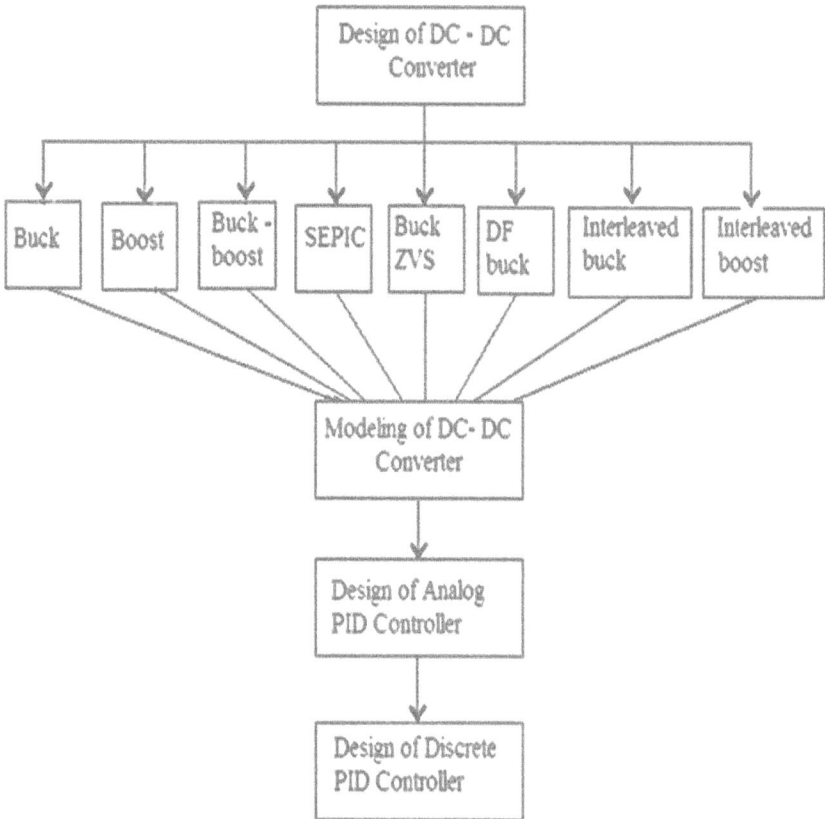

Figure 3.1 Overview of the controller for DC-DC converter.

A closed-loop reaction of a discrete controller is calculated by a little samples of the digital compensator. This principle is applied to programme a discrete PID controller to produce the preferred result. The discrete controller approach prevents errors associated with the time domain method of converting from the continuous to the discrete domain. Initially, an analogue PID controller is configured for a DC-DC converter using Ziegler Nicholas' second process. The continuous domain PID controller is then transformed to discrete domain. There are four: a) Impulse invariance b) Phase invariance c) Finite difference and d) bilinear transformation techniques for transforming continuous domain to discrete domain. Among these, the Tustin approach search the analogue controller output closer at sample periods and approximates to the analogue integration is better than other approaches. Since it is consistent with the configuration of all forms of filters and compensators, the suggested controller implements the bilinear transformation methodology. Furthermore, assured reliability is a desirable performance for any controller style.

The three main things to remember in the construction of a digital controller for a DC/DC converter are 1) To sample the error signal for converting the Analog input to Digital input 2) To correct the error signal by using wireless compensator 3) High-resolution DPWM signals are provided to achieve high precision in needed output voltage. To keep the system switching frequency high, high resolution DPWM signals are produced in this work. In order to attain the goal of the Controller, DC to DC converter switches are operated so that the fraction of the voltage output is to be made in identical with the reference voltage. In continuation with that the changes in load or input voltage should endure remain constant.

3.2 Design of DC-DC Converter

To change an uncontrolled DC voltage to a control one, DC to DC converters might be a substitute as switching mode regulators. These DC to DC converter operation has been divided into two modes those are: Continuous Conduction Mode and Discontinuous Conduction Mode. The resistive load output voltage of a converter is discontinuous and incorporates harmonics. An LC filter is typically used to reduce the ripple content. It is assumed that i) all of the components R, L, and C are perfect, ii) the circuit is in steady-state operation, and iii) constant current flows through the inductor. iv) The swapping interval is T, and the power electronic switch is shut for time interval DT and it is open for time interval DT (1-D) is a big capacitor with a fixed output voltage. The converter switching frequency has been determined by the values of the R (resistor) and C (capacitor). To determine the value of inductance (L), the ripple current (I) will be limited to 10% - 30% of the full load current, and to determine the value of capacitance, the ripple voltage (Vc) will be limited to 1% - 2% of the output voltage.

3.2.1 Design of Buck Converter

A DC to DC power converter is an electrical circuit that varies the level of voltage to a direct current source. It is critical in portable electronic devices such as laptop computers, and cellular phones, which rely on batteries for electricity. It also controls the voltage at the output. A DC input voltage source, intermediary electronics, and a load (resistive) that necessitates a power for separate DC voltage source are all part of a DC-DC converter.

Voltage step down and current step up converters are both referred to as buck converters. This is a DC/DC converter, and this is the easiest way to lower a DC supply's voltage. It is remarkably highly efficient (more than 90%).

The current begins to grow when turn ON the switch 'S', and the inductor responds by providing a voltage between its terminals. By counteracting the source voltage, this voltage reduces load voltage remaining in it. The inductor current changes at a slower rate over time, reducing the inductor voltage and boosting the load voltage. The inductor becomes a tremendous source of energy which is in the form of magnetic field when S (the switch) is closed. The net voltage at the load is always lower than the input voltage source because there is a potential drop across the inductor is remained opened.

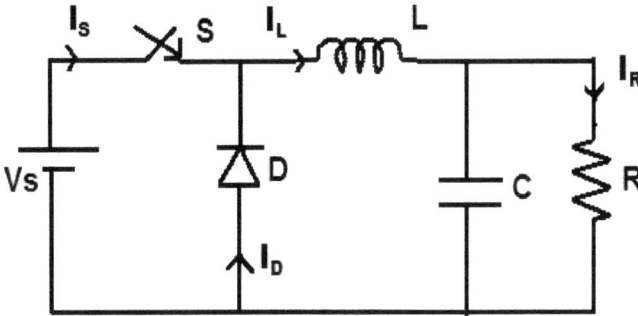

Figure 3.2 Conventional buck converter.

There are two modes of operation in a buck converter which are continuous and discontinuous conduction of operation. The current flows through an inductor not ever fails to zero during Continuous conduction mode (CCM). During the discontinuous conduction mode, the inductor current is zero, a portion of time the amount of load energy is too small. This kind of operation is called as discontinuous conduction mode.

The CCM is used by the Buck converter is depicted in Figure 3.2. The semiconductor switch (S), diode (D), inductor (L) and capacitor (C) make up the buck converter. R signifies the load resistance, while Vs represents the input voltage. The detailed design of buck converter is given below.

The buck converter's output voltage is always smaller than the input and is assumed by

$$d = \frac{V_O}{V_S} = \frac{T_{ON}}{T_S} \tag{1}$$

Where T_{ON} is the ON time period, and T_S is the total ON and OFFF time of the switch (S) and d is the duty cycle. The conducting period T_S is denoted by

$$T_S = \frac{\Delta I L V_S}{V_O(V_S-V_O)} \tag{2}$$

As ΔI is the ripple conductor through an inductor and the inductor(L) value may be calculated as

$$Lf_S\, \Delta I_L \;=\; V_S\, d(1-d) \tag{3}$$

The switching frequency is denoted by fs. The inductor (L) is chosen with the knowledge that the inductor (L) value determines the size of the ripple current flows through the capacitor (C) as well as the load current (R). As a result, a ripple current of 20% to 30% of the average output current is often anticipated for the design to ensure satisfactory converter performance. The allowable output ripple voltage, which is 1 to 2 percent of the output voltage Vc, is then used to determine the capacitor C. The ripple voltage across the capacitor (C) is calculated using the formula:

$$\Delta V_c = \frac{V_s d(1-d)}{8 LC\, f_s{}^2} \tag{4}$$

The calculated parameter of the converter as, the source voltage is 12V, switching frequency as 400 KHz, inductor (L) as 12 µH, capacitor as 10 µF, and 41.67% is the duty cycle of the PWM pulse hence the resultant as 5V, 2A, 10W as output voltage, current and power respectively.

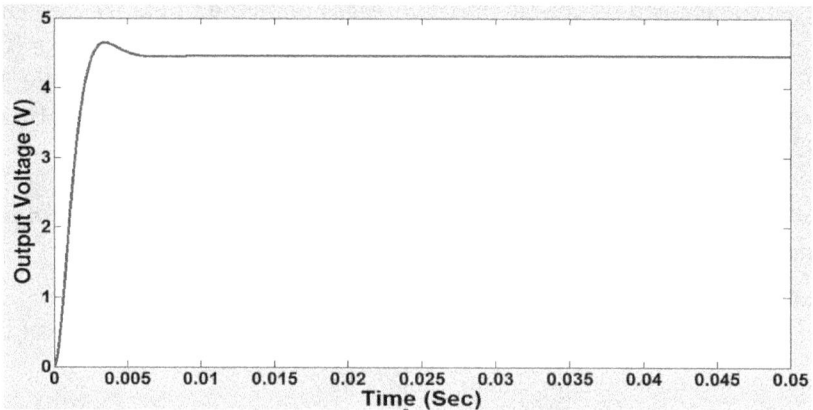

Figure 3.3 Open loop response of Buck converter for 5V output.

Figure 3.3 depicts the open loop response of Buck converter. In the obtained output, where the maximum overshoot and steady state error is found to be maximum as well as the voltage ripples are noticed which necessitates the design of closed loop control.

3.3 Modeling of DC/DC Converter

The method of Circuit Averaging or the State Space Averaging are used to model a DC-DC converter after it has been built. In this scenario, the modelling of DC to DC converter using the State Space Averaging method. It's a programme that allows you to simulate a switching converter as a continuous linear system. The actual filter corner frequency f_c must be lesser than the switching frequency f_s for State Space Averaging to work. A closed loop system's power stage is a non-linear system. Nonlinear system is notoriously tough to model and predict. As a result, it is better to estimate the nonlinear system to a linear system. The state space averaging method is employed in that scenario. A DC to DC converter in continuous current mode (CCM) has dual circuit states [1].

In the converter the power electronic switch S drives with the series of pulses with the switching frequency fs. The DC-DC converter's state vector is known as $x(t) = \begin{bmatrix} I_L \\ V_C \end{bmatrix}$, where I_L and V_c are the current through an inductor and potential drop across the capacitor. Through the specified duty ratio d(k) for the kth time, the technique is defined with the subsequent series of state space equations in continuous time domain. The pulse series with a continuous switching frequency(f_s) activates the switch S. The technique is characterized with the subsequent series of state space equations in continuous time domain for the specified duty cycle d(k) during the kth period, where IL is the current through an inductor. A pulse series by a fixed switching frequency f_s is used to activate the switch S. The DC-DC converter's state vector is as follows:

$$\left. \begin{array}{l} \dot{x}\,(t) = Ax(t) + BV_s(t), \;\; Sw = 1 \\ \dot{x}\,(t) = Ax(t) + BV_s(t), \;\; Sw = 0 \end{array} \right\} \tag{5}$$

where Sw = 1 signifies the switch(S) - ON status and Sw = 0 signifies the switch(S) - OFF status.

3.3.1 Modelling of Buck Converter

For modelling of converter, three techniques could be used

➢ Circuit averaging method

➢ Current inject technique
➢ State space averaging method

The latter methodology offers several advantages over the other two, including:

- A more compact formulation of equations
- Possibility of obtaining more transfer functions
- It is easier to get transfer functions.

As a result, state space averaging is employed to simulate the DC/DC converter in this suggested study [3].

The state space averaging methodology may be used to mimic the buck converter once it has been designed. This approach is unusual in that it may be used to design for a certain type of input, such as a step, ramp, or impulse function, while also taking into account the start circumstances. This approach is useful for approximating the dynamics at low frequencies while ignoring the discontinuous impact provided by switching. The power converter circuit's system equations are also required. Now we'll talk about state space analysis.

A pulse order with a continuous switching frequency f_S drives the switch S. A buck DC-DC converter's state vector is distinct as $x(t) = \begin{bmatrix} i_L \\ V_C \end{bmatrix}$, The current flows through an inductor is I_L, while the potential drop across the capacitor(C) is V_c. The arrangements are shown with the subsequent set of state space calculations in constant time domain for the specified duty cycle d(k) throughout the kth period:

The following set of constant time state-space calculations defines in the system:

$$\left.\begin{array}{l} \dot{x}\,(t) = Ax(t) + BV_S(t) \\ y(t) = Cx(t) + DV_S(t) \end{array}\right\} \tag{6}$$

The state variable vector is x(t), the source vector is V_S, and the state coefficient matrices are A, B, C, and D. The DC to DC converter's state model is generated and explored further down. Only the CCM (Continuous Conduction Mode) of action allows for high power densities. When Switch (S) – ON, Diode (D) – OFF, and vice versa, diode (D) and

MOSFET (S) are constantly in a complimentary condition. There are two alternative modes of operation, each with its own set of state equations.

Mode1: Switch (S) - ON and Diode (D) - OFF

During mode1, semiconductor switch (S) is turned ON for a time interval of $0 \leq t \leq T_{ON}$ and the diode D is in the off state and the corresponding circuit of the DC to DC converter is revealed in Figure 3.4.

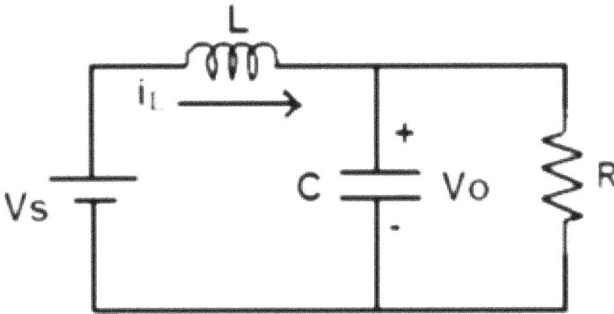

Figure 3.4 Equivalent circuit of Buck converter for Mode 1.

The input current, i_L completes its path through the inductor (L), the capacitor (C) and the load (R) [4]. During this interval supply terminal get connected across the load and the power deposited in the inductor. Since the diode gets reverse biased, the input energy is being fed to the inductor (L) and the load (R). The inductor and capacitor (C) combination act as a low pass filter and hence the voltage ripples are filtered out. The resonant frequency of the filter is selected to be considerably smaller than the frequency of the dc-dc converter in order to dominate essentially the presence of switching frequency ripple in the voltage output.

Applying Kirchoff's voltage and current laws, the dynamic equations governing the system during this interval are obtained as follows:

$$\left.\begin{array}{l} \dfrac{di_L}{dt} = \dfrac{(V_s - V_C)}{L} \\ \dfrac{dV_C}{dt} = \dfrac{i_L}{C} - \dfrac{V_C}{RC} \end{array}\right\} \qquad (7)$$

Here i_L and V_C are the state variables x_1 and x_2 respectively and hence the coefficient matrices for mode 1 are defined as,

$$\dot{x}(t) = A_1 x(t) + B_1 V_s(t) \tag{8}$$

where,

$$\left. \begin{aligned} A_1 &= \begin{bmatrix} 0 & \dfrac{-1}{L} \\ \dfrac{1}{C} & \dfrac{-1}{RC} \end{bmatrix} \\[2em] B_1 &= \begin{bmatrix} \dfrac{1}{L} \\ 0 \end{bmatrix} \end{aligned} \right\} \tag{9}$$

where, $\dot{x}(t) = \begin{bmatrix} x_1(t) \\ x_2(t) \end{bmatrix} = \begin{bmatrix} i_L(t) \\ V_o(t) \end{bmatrix}$ (10)

Mode2: S -turn OFF and D - turn ON

During mode 2, S is switched OFF and diode D is switched ON during the time interval $T_{ON} \le t \le T$. The inductor current i_L starts going to the inductor(L), capacitor and the Resistance load. Figure 3.5 depicts the buck converter's comparable circuit for this mode.

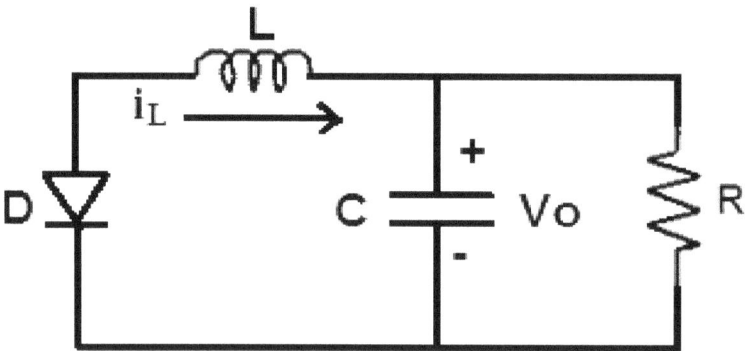

Figure 3.5 Equivalent circuit of buck converter for Mode 2.

$$\left.\begin{array}{l} \dfrac{di_L}{dt} = \dfrac{(-V_C)}{L} \\[2mm] \dfrac{dV_C}{dt} = \dfrac{i_L}{C} - \dfrac{V_C}{RC} \end{array}\right\} \tag{11}$$

Hence

$$\dot{x}(t) = A_2 x(t) + B_2 V_s(t) \tag{12}$$

where,

$$\left.\begin{array}{l} A_2 = \begin{bmatrix} 0 & \dfrac{-1}{L} \\[2mm] \dfrac{1}{C} & \dfrac{-1}{RC} \end{bmatrix} \\[6mm] B_2 = \begin{bmatrix} 0 \\ 0 \end{bmatrix} \end{array}\right\} \tag{13}$$

Here $x(t)$ is the state variable trajectory, A_1, B_1, A_2 and B_2 are the system matrices respectively. The two set equations can be replaced by a single correspondent set defined as follows,

$$\dot{x}(t) = [A][x] + [B][u] \tag{14}$$

The subjective means of real matrices characterizing the switching scheme provided by the subsequent equations are the A and B matrices.

$$\left.\begin{array}{l} A = dA_1 + (1 - d)A_2 \\[2mm] B = dB_1 + (1 - d)B_2 \end{array}\right\} \tag{15}$$

Hence,

$$\left.\begin{array}{l} A = \begin{bmatrix} 0 & \dfrac{d-1}{L} \\[2mm] \dfrac{1-d}{C} & \dfrac{-1}{RC} \end{bmatrix} \\[6mm] B = \begin{bmatrix} \dfrac{d}{L} \\[2mm] 0 \end{bmatrix} \end{array}\right\} \tag{16}$$

The output voltage $V_O(t)$ through the load(R) is stated as

$$V_o(t) = Y = [0 \ 1] \begin{bmatrix} I_L \\ V_C \end{bmatrix} + [0] \ V_S(t) \tag{17}$$

By relieving the standards of L and C in the state equations, the state coefficient matrices for the Buck converter is obtained follows,

$$\left.\begin{aligned} A &= \begin{bmatrix} 0 & -83333.33 \\ 100000 & -40000 \end{bmatrix} \\ B &= \begin{bmatrix} 34722.5 \\ 0 \end{bmatrix} \\ C &= [0 \ 1] \\ D &= [0] \end{aligned}\right\} \tag{18}$$

The buck converter's transfer function is subsequently derived, and the result is as follows:

$$tf = \frac{-7.276x10^{-12}S + 3.472x10^9}{S^2 + 40000S + 8.33x10^9} \tag{19}$$

3.4 Design of Analog PID Controller

In industry, the most widely used controller is the analogue PID controller. Three of the most significant proportional, integral, and derivative controller systems are included in the PID controller. The PID controller's transfer feature has

$$M(s) = K_P + \frac{K_I}{s} + K_D s \tag{20}$$

The letters K_P, K_I, and K_D stand for additive, integral, and derivative gain, respectively. Using the characteristic equation $(1 + G(s)H(s))$, the gain K_P, K_I, and K_D of the DC-DC converter were calculated. $G(s)$ is the DC-DC converter's transfer mechanism, and $H(s)$ is the unity input gain. There are various approaches for determining the controller's benefit. The Ziegler–Nichols method, for example, defines a basic mathematical technique for tuning PID controllers. The Routh - array system could be used to regulate the critical gain $(K_C)$and the resulting time of constant oscillation T_C. $K_P = K_C$ is the relative gain value. K_I and K_D can be found using the Ziegler–Nichols formulae. The K_P helps to shorten the rise time, the K_D to shorten the peak overshoot and settling time, and the K_I to remove the error during steady-state condition. As a result, controller gains boost the controller's dynamic response.

3.4.1 Design of Analog PID Controller for Buck Converter

The converter DC to DC whose output is controlled by a controller called PID (Proportional Integral Derivative). The PID controller examines the present error value, the sum of the error through a current time period, and the existing derivative of the signal error is to decide not only the amount of correction to apply, but also for how far [5].

A PID controller is a controller which has three kinds of control, and it is straightforward to understand. The letters P denotes proportional, I signifies integral, and D represents derivative. The simplest basic version of PID controller's transfer function,

$$C(S) = K_P + \frac{K_I}{S} + K_D S \tag{21}$$

K_P is the Proportionality constant
K_I is the Integral constant
K_D is the gain of Derivative

In a closed loop unity feedback system, let's use analogue PID controller C(S). The Analog PID controller receives the tracking error 'e.' The relative gain (K_P) times the amount of the error, in addition of the derivative gain (K_D) times the derivative of the error, with that the integral constant (K_I) times the integral of the error, equals the control output 'u' from the controller.

$$u = K_P e + K_I \int e \, dt + K_D \frac{de}{dt} \tag{22}$$

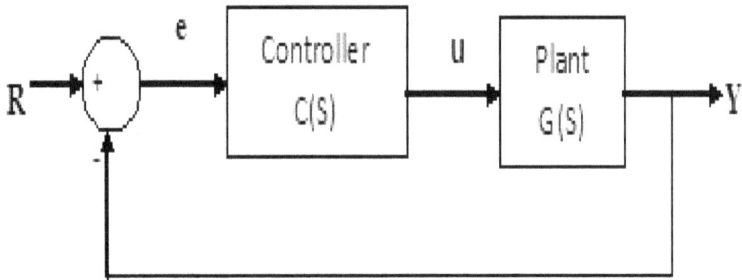

Figure 3.6 Basic structure of analog PID controller.

PID controllers are commonly utilized in industry owing to the ease of use and high performance. Closed loop industrial processes employ it in more than 90% of the time. Other sophisticated controllers are more sophisticated, but they only deliver minor improvements, whereas the PID controller can be modified by an operator with no prior experience with controls. This is especially important when the system is unfamiliar.

The four basic characteristics of a controller triggered by step input are:

➢ Rise time: The first time the plant output to reach 90% of the intended level is known as rise time.
➢ Settling time: The total time taken for the plant to attain its steady state is called settling time.
➢ Steady state error: The error between the required output and the steady state output is called steady state error.
➢ Peak overshoot: The maximum level exceeds the steady-state value is called overshoot.

Table 3.1 represents the effect of the controller parameters

Table 3.1 Effect of controller parameters

Parameters	Rise-time	Maximum Overshoot	Settling-time	Steady-state-error
K_P	Reduce	Increase	NDT	Reduce
K_I	Reduce	Increase	Increase	Eliminate
K_D	NDT	Reduce	Reduce	NDT

** NDT : No Definite Trend

1. Determine which aspects of the system need to be enhanced when constructing a PID controller.
2. Reduce the rising time by using K_P.
3. Reduce overshoot and settling time using K_D.
4. To reduce the steady-state inaccuracy, use K_I.

Ziegler and Nichols developed methods for estimating standards of the K_P (proportional gain), T_I (integral time constant), and T_D (derivative time constant) based on the transient response characteristics of a specified plant to implement PID controllers.

In this work Ziegler–Nichols second method is adopted to find PID controller parameters namely K_P, T_I and T_D values. Set $T_I = 0$ and $T_D = 0$ in the second technique. Increasing K_P from 0 to a critical value K_{cr} at which the output initially displays persistent fluctuations by means of the proportional control act. As a result, the critical gain K_{cr} and the accompanying period P_{cr} are found experimentally.

Table 3.2 Ziegler–Nichols fine-tuning rule centered on critical gain (K_{cr}) and critical period (P_{cr})

Types of Controller	K_P	T_I	T_D
P	$0.5K_{cr}$	∞	0
PI	$0.45\ K_{cr}$	$\dfrac{1}{1.2P_{cr}}$	0
PID	$0.6\ K_{cr}$	$0.5\ P_{cr}$	$0.125\ P_{cr}$

The PID controller adjusted by second way of Ziegler–Nichols rules contributes

$$
\begin{aligned}
C(S) &= K_P \left(1 + \frac{1}{T_I S} + T_D S\right)\\
&= 0.6K_{cr}\left(1 + \frac{1}{0.5P_{cr}} + 0.125P_{cr}\right)\\
&= 0.075K_{cr}P_{cr}\left(S + \frac{4}{P_{cr}}\right)^2
\end{aligned}
\qquad (23)
$$

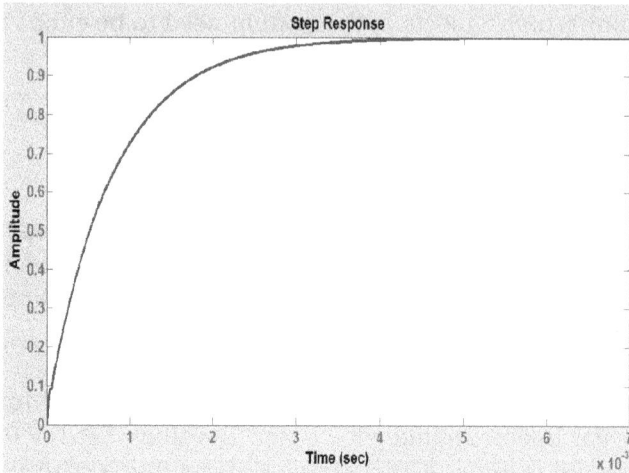

Figure 3.7 Analog PID controller step response for buck converter.

Thus, the PID controller, whose pole at the origin and its double zeros at $= \dfrac{-4}{P_{cr}}$.

Use the root – locus method to find the critical gain K_{cr} and the frequency of the sustained oscillations ω_{cr}, where $P_{cr} = \dfrac{2\pi}{\omega_{cr}}$. These values can be found from the crossing points of the root – locus branches with the $j\omega$ axis.

In this work, PID controller parameters of the buck converter is

K_{cr} = 0.2, ω_{cr} = 108637 rad/sec, $P_{cr} = \dfrac{2\pi}{\omega_{cr}} = 5.784 \; x \; 10^{-4}$

$K_P = K_{cr} = 0.2$

$T_D = 0.125 \; P_{cr} = 7.23 \; x \; 10^{-6}$

$T_I = 0.5 \; P_{cr} = 2.89 \; x \; 10^{-5}$

$C(S) = K_P \left(1 + \dfrac{1}{T_I S} + T_D S \right)$

$C(S) = 0.2 \left(1 + \dfrac{1}{2.89 \; x \; 10^{-5} S+} + 7.23 \; x \; 10^{-6} S \right)$

3.5 Design of Discrete PID Controller

The following equation represents the general shape of the traditional configurations of a digital PID compensator:

p(m) = [p(m-1) + a f(m) + b f(m-1) + c f(m-2)]   (24)

where, p(m) is the control signal ratio, f(m) is the feedback error signal, and a, b, c are the compensator frequency response shaping coefficients. Typically, the coefficients are derived by either a direct or indirect process. The controller we suggest is of the indirect form. The
PID compensator is built in continuous time and then transformed to
discrete time using the bilinear transformation technique. Approach
the following protocol to create the digital P(z) for the operation.

Determine the dynamic characteristics of the operation using open –
loop experiments.

1) Using the Ziegler–Nichols system, tuning rules were established to
obtain the K_C, T_I, and T_D of the PID controller.
2) Using a bilinear transformation technique and a sampling time of
T = 1S, convert the continuous domain parameters to discrete domain
parameters, and the digital PID control will look like this [2]:

$$M(z) = K_C \left[1 + \frac{T}{2T_I}\left(\frac{z+1}{z-1}\right) + \frac{T_D}{T}\left(\frac{z-1}{z}\right)\right] \tag{25}$$

$$M(z) = \left[\frac{\left(K_P + K_I\frac{T}{2} + \frac{2K_D}{T}\right)z^2 + \left(K_IT_s + \frac{4K_D}{T}\right)}{z(z-1)}\right] \tag{26}$$

$$M(z) = \frac{(z-z1)(z-z2)}{z(z-1)} \tag{27}$$

A digital PID controller is made up of two zeros, and it normally has
two zeros in addition to the pole at (1, 0) in the z-plane, which corre-
sponds to the pole at the origin in the continuous time domain. The
poles are used to remove steady-state error, while the zeros are used
to improve phase response at high frequencies. The following para-
graph explains the design of discrete PID controller for buck converter and its performance.

3.5.1 Design of Discrete Controller for Buck Converter

Digital controller for power converters has many advantages over analog controller. It has more flexible, integratable, reliable, cost-effective, and less susceptible to noise and drift. Digital controller is more

common in the case of high-frequency converter. Because of these benefits, digital controller is desirable for all types of power converters.

In the continuous-time domain controller, the Trapezoidal technique is known as the Tustin technique or Bilinear – Transformation technique is applied to transfer equation into the discrete-time domain. By sampling at regular time periods and approximation to the analogue integration, the Tustin technique follows the analogue controller output more precisely than other approaches. The following is a trapezoidal approximation:

The value of the integral of t = (N+1) T is the same to the value at NT plus the area added from NT to (N+1)T if m(t) equals the integral of e(t).

$$M[(N+1)T] = u(NT) + \int_{NT}^{(N+1)T} e(\tau)d\tau \tag{28}$$

By means of Trapezoidal rule, e(t) is the area curve from t = NT to t = (N+1)T is approximated as

$$\frac{e[(N+1)T] + e(NT)}{2} \ x \ T \tag{29}$$

Therefore

$$M\{(N+1)T = n(NT) + \frac{T}{2} \ \{e[(N+1)T] + e(NT)\} \tag{30}$$

Taking the z-transform of (32) then

$$zN(z) = N(z) + \frac{1}{2}[zE(z) + E(z)] \tag{31}$$

Thus $\quad \dfrac{M(z)}{E(z)} = \dfrac{T}{2}\left[\dfrac{z+1}{z-1}\right]$ $\tag{32}$

As a result, eq. (32) is the discrete Integrator transfer function. When the derivative of e(t) at interval(t) = NT is m(NT), we have a trapezoidal approximation to differentiation.

$$m(NT) \cong \frac{e(NT) - e[(N-1)T]}{T} \tag{33}$$

Taking z – transform of (33)

$$\frac{M(z)}{E(z)} = \frac{(z-1)}{Tz} \tag{34}$$

The digital PID controller transfer function [2] becomes

$$G(z) = \frac{V(z)}{E(z)} = \left[K_P + K_I \frac{Ts}{2} \frac{z+1}{z-1} + K_D \frac{z-1}{Tz} \right] \tag{35}$$

$$V(z) = \left[\frac{\left(K_P + K_I \frac{Ts}{2} + \frac{2Kd}{Ts} \right)z^2 + \left(K_I T_s + \frac{4K_D}{T_s} \right)}{z(z-1)} \right] E(z) \tag{36}$$

Now substitute eq. (36) in the intended buck DC-DC converter, then the discrete controlled buck converter is given as:

$$V(z) = \frac{3.194z^2 - 5.77521z + 2.6942}{z\,(z-1)} \tag{37}$$

$$V(z) = \frac{(z-0.94922)(z-0.91743)}{z(z-1)}$$

The stability of the digital controller has been verified through the root locus method is illustrated in Figure 3.8.

The Root locus response is given in Figure 3.8 to examine the stability of the regulated discrete buck converter. The attributes of stability include:

1. Any closed loop Pole outside the unit circle represents destabilizes in the system.
2. The system becomes critically stable if a simple pole is located at z = -1.
3. The system come to be unstable if there are many closed – loop poles on the unit circle.
4. Because closed loop zeros have no effect on stability, they can be placed wherever in the z plane.

The planned buck converter's discrete PID controller transfer function is

$$V(z) = \frac{3.0942z^2 - 5.77523z + 2.6941}{z^2 - z} \tag{38}$$

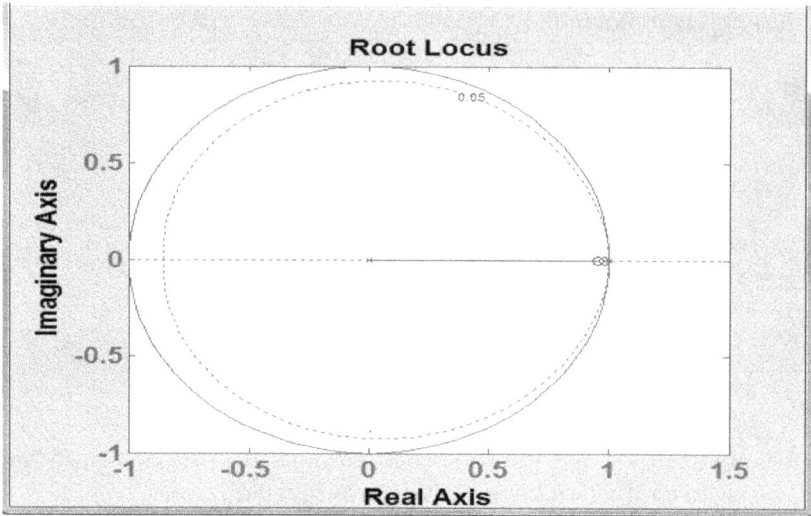

Figure 3.8 The suggested discrete PID controller for Buck converter's root locus response.

In the transfer function equation (38), a root locus plot has been created. The poles are clearly situated neither beyond the unit circle nor at -1, as shown by the root locus map. There haven't been any multiple poles. All poles are in the right half of the z-plane, meeting the transfer function frame's stability criteria for our proposed controller.

3.6 Design of PWM

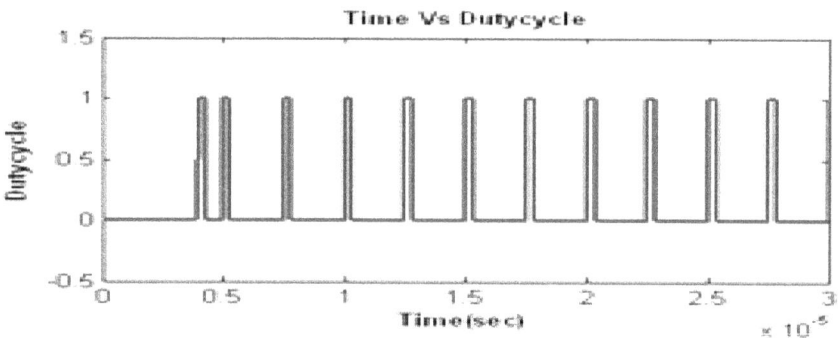

Figure 3.9 PWM pulse output.

The discrete time integral compensator reduces the error and transmits the command input to the switch as pulses, allowing the signal output to track the reference input signal. The compensator's signal output is a digital signal that is quantized and saturated into an analogue signal.

The corrected error signal from the digital compensator is associated contrary to the high-frequency (carrier) ramp signal is used to obtain the duty cycle pulse for the converter. The carrier signal frequency is switching frequency (400 KHz) of the converter. The output signal of the digital pulse width modulation is given in Figure 3.9.

Table 3.3 shows a comparison of the controllers centered on the state-space system and which is recognized in converter model. The discrete PID controllers that are takeout based on the proof of identity give superior lively outcomes in non-open loop response than the additional controllers, according to the findings. The suggested approach has several benefits, including the detail that it is performed in the time-domain (and therefore avoids many mistakes caused by the s to z-transformation) and it has not need in a trial and error technique. Table 3.4 shows the fluctuations in various component characteristics as well as the controller signal output response. The controller stability has been demonstrated by Table 3.4 and simulation results, which show that fluctuations in component and input voltage have no effect on the output voltage.

Table 3.3 Evaluation of the various controllers for Buck converter

Controller	Settling Time (ms)	Maximum-Overshoot (%)	Rise Time (ms)	Steady State Error (V)	Ripple Voltage (V)
Digital PID	1	0	1	0.00398	0
Digital PI	12	5	2	0.01	0
Analog PID	25	4.3	11	0.0001	0.05
Analog PI	22	4.5	11	0.02	0.09
Sliding mode	1.2	2.6	0.08	0.001	0.001

Table 3.4 The performance of the discrete PID controller as a function of different factors

Component Variation in Output Response					Input Voltage Variations and Output Response		
R in Ohm	L in Micro Hendry	C in Micro Farad	Reference signal in volts	Output signal in volts	Input Voltage in volts	Reference Voltage in volts	Output Voltage in volts
3	-	-	6	6	12	6	6
6	12	-	6	6	15	6	6
6	15	14	6	6	10	6	6
26	14	12	6	6	8	6	6
35	14	14	6	6	18	6	6
50	18	16	6	6	20	6	6

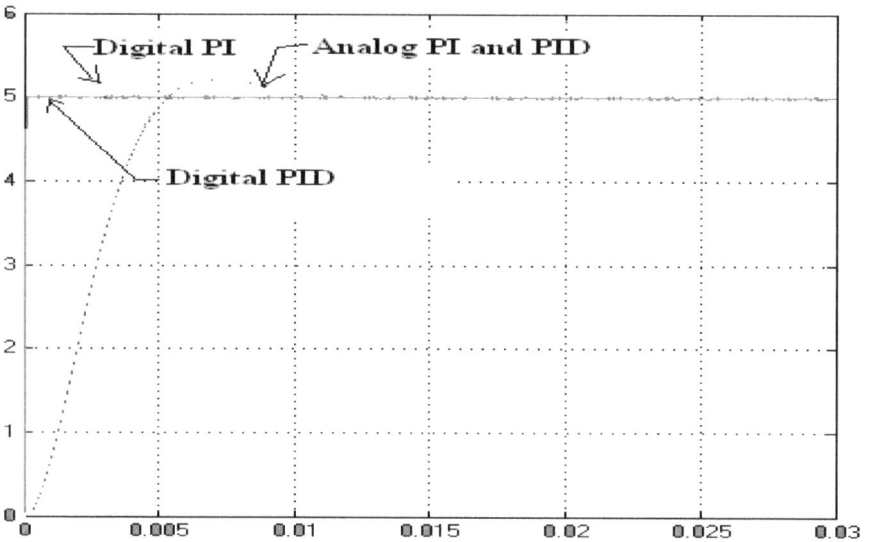

Figure 3.10 Analog and discrete PID, PI controller output voltage response for discrete controlled buck converter.

The output signal response of the analog and discrete controllers is shown in Figure 3.10. Analog controllers have a long rising & settling time, and peak overshoot as seen in the diagram. Although the discrete PI controller has a faster increase and decrease time than analogue controllers, it is the maximum overshoot of various controllers.

From Figure 3.10, one can understand that there is zero peak over-shoot, steady state error, ripple voltage, rising and settling time in a digital PID controller. This proves that the digital PID controller improves the dynamic performance of the buck converter.

3.7 Conclusion

This chapter explains how to build and construct a Discrete PID controlled Buck converter using a basic way. The proportional(P), integral(I), and differential(D) gains of a controller PID may be chosen built on the value of input error signal to increase the Buck converter's transient responsiveness greatly. The discrete PID controller is realized in a buck converter and its transient properties are tested in an experimental (hardware) way also. To demonstrate the enhanced non static response under nonlinear control for 12V/5V point of load Buck converter whose MATLAB simulation results and hardware experimental results are all provided in this chapter.

References

1. Alkrunz, Md., and Irfan, Y. (2015) Design of discrete time controllers for the DC-DC boost converter. *Journal of Science*, **20**(1), 203-211.
2. Vijayalakshmi, S., and Sree Renga Raja, T. (2014) Time domain based digital PWM controller for DC-DC converter. *Automatika*, **55**(4), 434-445.
3. Chander, S., Agarwal, P., and Gupta, I. (2011) Auto-tuned, Discrete PID Controller for DC-DC Converter for Fast Transient Response. *IEEE ICEMS Conference*, pp. 1-4.
4. Shenbagalakshmi, R., Vijayalakshmi, S., and Geetha. K. (2020) Design and analysis of discrete PID controller for Luo converter. *International Journal of Power Electronics*, **11**(3), 283-298.
5. Vijayalakshmi, S., and Sree Renga Raja, T. (2014) Time domain based digital controller for buck-boost converter. *Journal of Electrical Engineering Technology*, **9**(5), 1551-1561.

4

Intelligent Voltage Control based Energy Saving Scheme for Space Vector Modulation controlled Induction Motor Drives

Sowmmiya U.[1] and Jamuna V.[2]

[1]*Dept of Electrical & Electronics Engineering, SRM Institute of Science & Technology, Kattankulathur, India*
[2]*Dept of Electrical & Electronics Engineering, Jerusalem College of Engineering, Chennai, India*

Abstract

Induction motors are mainly used in industries and are often operated at no-load or partial-load condition and the motor is always connected to the mains causing rated iron loss. These losses are unutilized form of few primary energy sources. In induction motors if the terminal voltage is reduced by using a switching device, some electrical energy might be saved. Nowadays, voltage controllers play a vital role as soft starters for motors and at times as energy savers with reduction in the flux level. The Space Vector Modulation (SVM) inverters employed for the reported work, reduces the THD and improves the fundamental voltage. In this chapter, the three-phase induction motor model is detailed and the terminal voltage of the same is controlled using Space Vector Modulated Inverter. The control scheme to save energy from three phase induction motor drive is developed. From the work described in this chapter, it is evident that much energy savings can be affected during no-load and the saving decreases with the increase in the load.

4.1 Introduction

The demand for energy and the energy saving concepts has always been very important in the context of improvement in efficiency. Manufacturing industries, who consume a huge amount of energy has to have a watch on this issue and make their final product more competitive, cost-effective and energy saving. The motor drive systems installed in cities consume over half of all electricity and it can be computed to about 75% of all electricity in an industrial plant.

Thus, in industries it is necessary to bench mark the unit consumption per ton of final product output. AC variable speed drive is one of the many well known solutions to achieve this goal. The growing popularity of AC drives depends primarily on the control of the speed of induction motor which is the most needed segment of the motor control sector. Three phase induction machine is widely employed in industries owing to its simple construction, reliable operation, lightness and cheapness. For the right applications, when right motors are employed, energy can be saved. When the operation of induction machine is altered, the saving of energy is questioned. Induction motors are inherently very efficient when fully loaded. The efficiency will reduce during half loaded condition and thus demands improvement of efficiency through the advent of power electronics. The efficiency of induction machine can be improved when the voltage reduces considerably. In general, the energy saving can be minimal, and the best saving can be affected on high iron loss motor system. When payback periods are a consideration, the savings is a big concern and the power consumed will not create a big impact.

4.2 Modelling and System Description

In recent years, the usage of two stage converters for motor applications is evolving greatly with the intention of size reduction, improvement in efficiency and lifetime and cost reduction. The most near sinusoidal voltage and current devoid of low-order harmonics and UPF can be achieved. In two stage converters, during DC-AC conversion, modulation techniques are employed to obtain sinusoidal output voltage. The DC-link voltage is maintained constant with the help of large electrolytic capacitor. In this chapter, the work involves Space Vector Modulation (SVM) with 6 fixed active voltage vectors and 2 zero vectors for identifying the required rotating voltage vector.

4.2.1 AC-DC-AC Converter

A Three phase AC supply with 50 Hz fundamental frequency is given to the front-end diode ridge rectifier. It is a combination of six diodes which conduct on the 120-degree conduction mode. It converts the three phase AC supply into uncontrollable DC supply. The inductor and capacitor are employed to decrease the ripple content in DC output voltage. The inverter is a combination of six IGBTs. Here, DC supply is converted in to three phase AC supply. The block diagram of two stage converter with front end diode rectifier is shown in Figure 4.1. The output terminal of the inverter is connected to the motor. The speed is sensed and feedback to the v/f control unit. Space vector modulation controller calculates the sector timing and gives the signal to the gates of the IGBT.

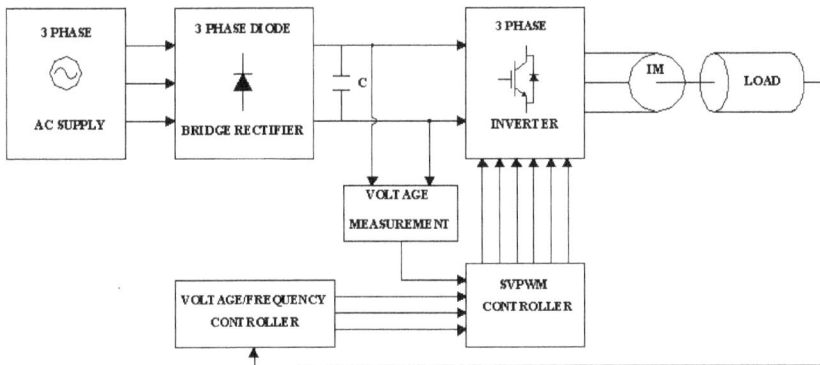

Figure 4.1 SVM controlled AC-DC-AC converter.

4.2.2 Space Vector PWM

Space Vector Modulation (SVM) assumes the inverter as a single unit with three separate push pull driver stages. The circuit representation of a three-phase voltage source inverter (VSI) is given in Figure 4.2. S_1 - S_6 are the six switches controlled by the switching variables a, a′, b, b′, c and c′. When a switch in the upper half is turned on then the corresponding switch in the lower half is switched off. Thus, the on and off states of the upper switches can be involved to identify the output voltage. The SVM presumes the sinusoidal voltage as a fixed amplitude vector revolving at constant frequency.

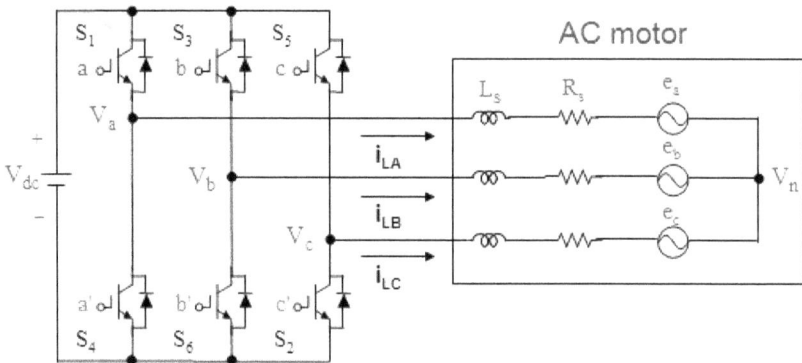

Figure 4.2 Three phase PWM inverter.

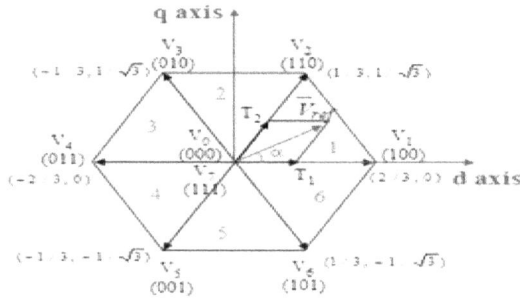

Figure 4.3 Basic switching vectors and sectors.

The three phase voltage vector is transformed into two phase voltage vector in the stationary d-q frame. The vectors (V_1 to V_6) splits the entire plane into 6 sectors of 60 degrees and V_{ref} is generated with 2 adjacent non-zero vectors and 2 zero vectors. The line-to-line voltage vector and line-to-neutral voltage vector are given by equation (1) and (2).

$$\begin{bmatrix} V_{ab} \\ V_{bc} \\ V_{ca} \end{bmatrix} = V_{dc} \begin{bmatrix} 1 & -1 & 0 \\ 0 & 1 & -1 \\ -1 & 0 & 1 \end{bmatrix} \begin{bmatrix} a \\ b \\ c \end{bmatrix}$$

(1)

$$\begin{bmatrix} V_{an} \\ V_{bn} \\ V_{cn} \end{bmatrix} = \frac{1}{3} V_{dc} \begin{bmatrix} 2 & -1 & -1 \\ -1 & 2 & -1 \\ -1 & -1 & 2 \end{bmatrix} \begin{bmatrix} a \\ b \\ c \end{bmatrix}$$

(2)

The reference vector is produced using SVM when switching takes place between two nearest active vectors and zero vectors. Figure 4.3 depicts the location of space and reference vector in the first sector.

4.2.3 Modulation Algorithm

In space vector modulation algorithm, the choice of the zero vector, sequencing of the vectors, splitting of the duty cycles of the vectors without incorporating any additional commutations are merits. The various types of SVM techniques are:

- Right aligned sequence (SVM$_1$)
- Symmetric sequence (SVM$_2$)
- Alternating zero vector sequence (SVM$_3$)
- High current not switched sequence (SVM$_4$) – Most commonly used

The high current not switched sequence scheme works on the fact that the switching losses are proportional to the highest current magnitude of the switched thereby avoiding the inverter leg carrying the highest instantaneous current. With the use of only one zero vector, V_7(ppp) or V_0(nnn) within a given sector one of the legs need not be switched at all, as shown in Figure 4.4.

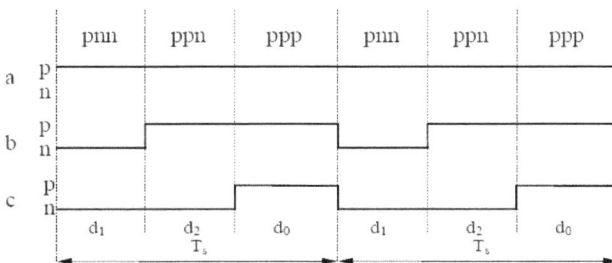

Figure 4.4 Phase gating signal of SVM 4.

The execution of the algorithm happens with the determination of V_d, V_q, V_{ref}, angle (α), time duration T_1, T_2, T_0 and the switching time of each transistor $(S_1$ to $S_6)$ using the following equations from (3) – (11).

$$V_d = V_{an} - V_{bn}.\cos60 - V_{cn}.\cos60 \tag{3}$$

$$V_q = 0 + V_{bn}.\cos30 - V_{cn}.\cos30 \tag{4}$$

$$\alpha = \tan^{-1}(v_q/v_d) \tag{5}$$

Switching time duration at any sector

$$\int_0^{T1} V_{ref} = \int_0^{T1} \bar{v}\,1 dt + \int_0^{T1+T2} \bar{v}2 dt + \int_{T1+T2}^{T2} \bar{v}\,0$$

$$T_1 = T_z.a.\frac{(\sin\frac{\pi}{3} - \alpha)}{\sin\left(\frac{\pi}{3}\right)} \tag{6}$$

$$T_2 = T_z.a.\frac{\sin(\alpha)}{\sin\left(\frac{\pi}{3}\right)} \tag{7}$$

$$T_0 = T_z - (T_1 + T_2)\ \left(\text{where, } T_z = \frac{1}{f} \text{ and } a = \frac{|V_{ref}|}{\frac{2}{3}V_{dc}}\right) \tag{8}$$

Switching time duration at any sector

$$T_1 = \frac{\sqrt{3.T_z.|V_{ref}|}}{V_{dc}}\left(\sin\left(\frac{\pi}{3} - \alpha + \frac{n-1}{3}\pi\right)\right)$$

$$= \frac{\sqrt{3.T_z.|V_{ref}|}}{V_{dc}}\left(\sin\frac{n}{3}\pi - \alpha\right)$$

$$= \frac{\sqrt{3.T_z.|V_{ref}|}}{V_{dc}}\left(\sin\frac{n}{3}\pi\cos\alpha - \cos\frac{n}{3}\pi\sin\alpha\right) \tag{9}$$

$$T_2 = \frac{\sqrt{3.T_z.|V_{ref}|}}{V_{dc}}\left(\sin\left(\alpha - \frac{n-1}{3}\pi\right)\right)$$

$$= \frac{\sqrt{3.T_z.|V_{ref}|}}{V_{dc}}\left(-\cos\alpha.\sin\frac{n-1}{3}\pi + \sin\alpha.\cos\frac{n-1}{3}\pi\right) \tag{10}$$

$$T_0 = T_z - T_1 - T_2, \left(\begin{matrix} where, n = 1 \text{ through } 6 (\text{that is, sector1 to 6}) \\ 0 \leq \alpha \leq 60° \end{matrix} \right) \quad (11)$$

4.2.4 Modeling of Three-Phase Induction Motor

An induction motor (IM) is a type of asynchronous AC motor and is designed using Krause's theory with the inputs as three-phase voltages at their fundamental frequency. The dynamic equivalent model of induction machine is shown in Figure 4.5.

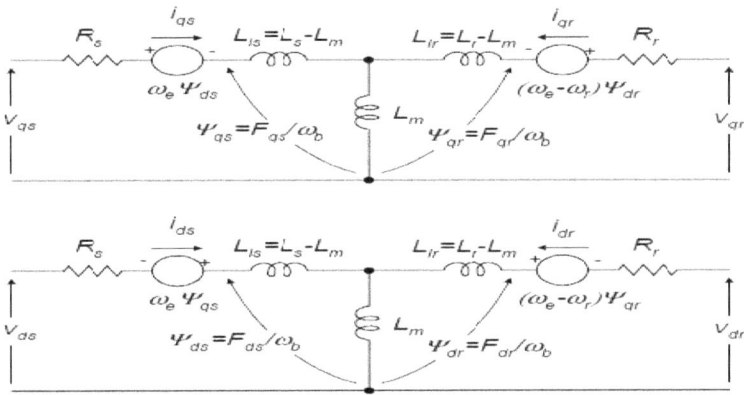

Figure 4.5 Dynamic or d-q model of induction machine.

The modeling equations depending on the flux linkage form are shown from (12)-(23) and the simulation implementation of the same is shown in Figure 4.6.

$$dF_{qs}/dt = \omega_b[v_{qs} - (\omega_e/\omega_b)*f_{ds} + R_s/x_{ls}(f_{mq}+f_{qs})] \quad (12)$$

$$dF_{ds}/dt = \omega_b[v_{ds} + (\omega_e/\omega_b)*f_{qs} + R_s/x_{ls}(f_{md}+f_{ds})] \quad (13)$$

$$dF_{qr}/dt = \omega_b[v_{qr} + (\omega_e-\omega_r/\omega_b)*f_{dr} + R_r/x_{lr}(f_{mq}-f_{qr})] \quad (14)$$

$$dF_{dr}/dt = \omega_b[v_{dr} + (\omega_e-\omega_r/\omega_b)*f_{qr} + R_r/x_{lr}(f_{md}-f_{dr})] \quad (15)$$

$$F_{mq} = x_{ml}^*[f_{qs}/x_{ls} + f_{qr}/x_{lr}] \quad (16)$$

$$F_{md} = x_{ml}^*[f_{ds}/x_{ls} + f_{dr}/x_{lr}] \quad (17)$$

$$I_{qS}=1/x_{ls}[f_{qs}-f_{mq}] \tag{18}$$

$$I_{dS}=1/x_{ls}[f_{ds}-f_{md}] \tag{19}$$

$$I_{qr}=1/x_{lr}[f_{qr}-f_{mq}] \tag{20}$$

$$I_{dr}=1/x_{lr}[f_{qr}-f_{md}] \tag{21}$$

$$T_e=[3/2]*[P/2]*[1/\omega_b]*[f_{ds}i_{qs}-f_{qs}i_{ds}] \tag{22}$$

$$T_e-T_L=[J]*[2/P]*[d\omega_r/dt] \tag{23}$$

The above equations are rearranged in the state-space form with state variables as x=[f_{qs} f_{dr} f_{qr} f_{dr} w_r]^T. The modeling equations of squirrel cage induction motor in state-space are given by equations (24)-(28).

$$dF_{qs}/dt=\omega_b[v_{qs}-(\omega_e/\omega_b)*f_{ds}+R_s/x_{ls}(X_{ml}^*/X_{lr}*f_{qr}+(x_{ml}^*/x_{ls}-1)f_{qs})] \tag{24}$$

$$dF_{ds}/dt=\omega_b[v_{ds}-(\omega_e/\omega_b)*f_{qs}+R_s/x_{ls}(X_{ml}^*/X_{lr}*f_{dr}+(x_{ml}^*/x_{ls}-1)f_{ds})] \tag{25}$$

$$dF_{qr}/dt=\omega_b[v_{qr}+(\omega_e-\omega_r/\omega_b)*f_{dr}+R_r/x_{lr}(X_{ml}^*/X_{lr}*f_{qr}+(x_{ml}^*/x_{ls}-1)f_{qr})] \tag{26}$$

$$dF_{dr}/dt=\omega_b[v_{dr}+(\omega_e-\omega_r/\omega_b)*f_{qr}+R_r/x_{lr}(X_{ml}^*/X_{lr}*f_{qr}+(x_{ml}^*/x_{ls}-1)f_{qr})] \tag{27}$$

$$dw_r/dr=(P/2J)(T_e-T_L) \tag{28}$$

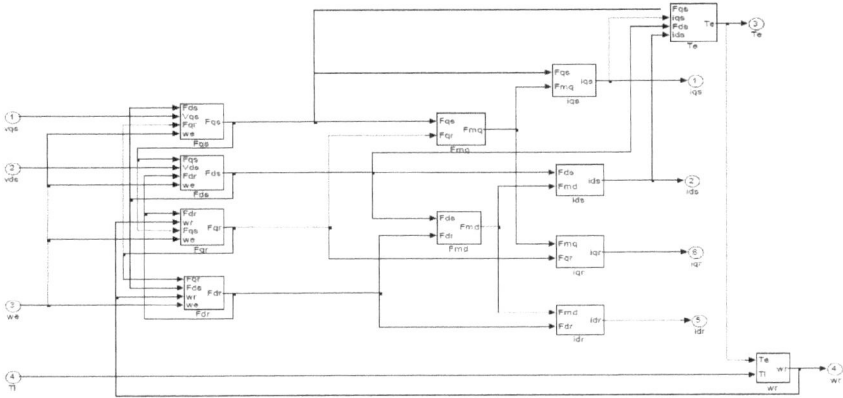

Figure 4.6 Induction machine d-q model block implementation using Simulink.

4.2.5 Energy Conservation Scheme

In industrial complexes, the induction motors may run at no or low partial load and irrespective of the load condition, these loads always stays connected to the mains. Effective energy conservation will improve the energy scenario and the overall efficiency. In the recent years by introducing a power electronic controller between the source and the load effectively improves the energy performance. The various energy conservation methods are by using soft starter, auto transformer, and voltage controllers. The performance of an induction motor can be improved by varying the load in accordance to the variation in supply voltage. The work reported in this chapter involves the supply of AC voltage to the induction motor through the inverter. The circuit is simulated for different loading condition of the motor and the waveforms of power factor, speed and slip are observed. By using suitable control technique, the applied voltage is controlled. At no load condition, the induction motor voltage is varied from 20% to 100% of the rated value of voltage and the energy saving is calculated.

Tests Performed on Induction Motor

No load test and blocked rotor test are conducted for 4KW, 230V, 10.5A, 1440rpm three phase induction motor, and the parameters are given in Table 4.1.

Table 4.1 Three phase induction motor parameters

R_r=0.39Ω	R_s=0.19 Ω
L_{ls}=0.21e-3H	L_{lr}=0.6e-3H
L_m=4e-3H	F_b=50Hz
P=4	J=0.0226Kgm^2
L_r=L_{lr}+L_m	T_r=L_r/R_r
X_{lr}=w_b*L_{lr}	X_{ls}=w_b*L_{ls}
X_m=w_b*L_m	Xm_{star}=1/(1/X_{ls}+1/X_m+1/X_{lr})

Open Loop Model

The open loop operation of three phase induction motor under no load condition is shown in Figure 4.7. Three phase voltages are given to the induction motor and the simulation is done for no load and partial load conditions. The various output waveforms for losses, slip and power factor are observed. By using the SVM technique, the applied voltage is controlled. At no load condition, the voltage of the motor is changed from 20% to 100% of the rated value of voltage and the savings in energy is calculated.

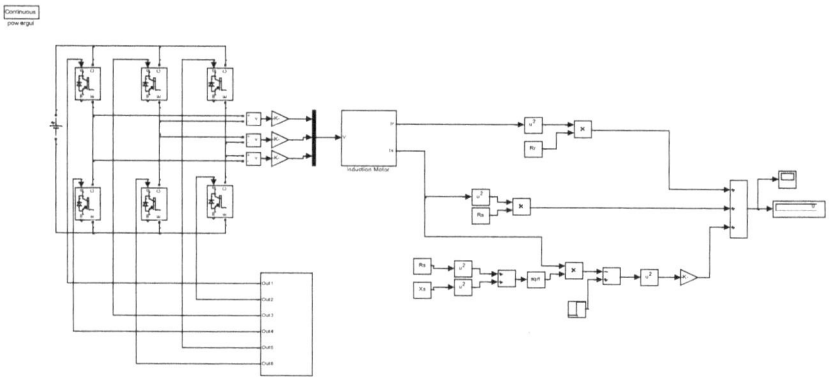

Figure 4.7 Open loop model of three phase induction motor under no load.

Figure 4.8 shows Speed Vs Time graph of the three phase induction motor running under no load condition with the maximum voltage applied to the motor terminals. It is noted that the maximum speed is 1497rpm.

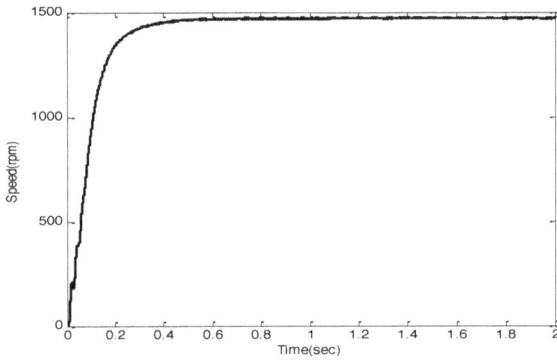

Figure 4.8 Speed vs time (Voltage = 230V).

The power factor of the machine for various voltages is calculated. For full voltage, the power factor is 0.32 and for 20% of the rated voltage the power factor is 0.77. Thus, the power factor is poor at full load condition, and it gets improved as the voltage decreases. The losses in the induction machine are calculated at various input voltages. The losses at full load and no load are found to be 2512W and 1061W respectively.

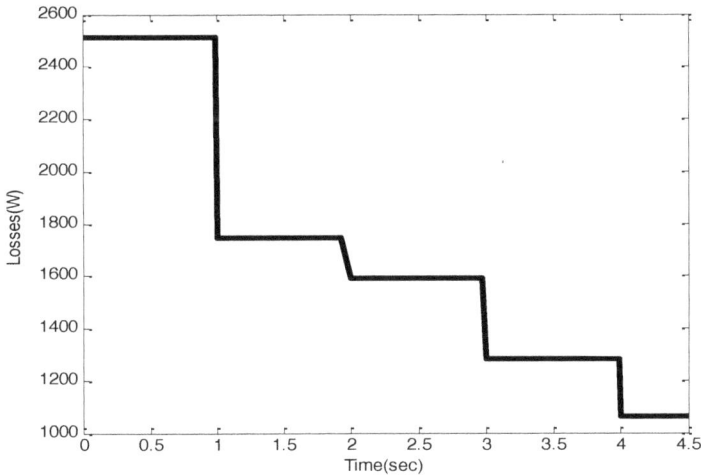

Figure 4.9 Losses vs time.

Figure 4.9 shows the graph between time and losses. It is observed that, at reduced voltage, the losses are reduced. Energy saving can be calculated from the losses.

Table 4.2 Parameters at no load condition

Voltage (V_{ph})	Speed (rpm)	Power Factor	Losses (W)	% Energy Saving	% Slip
230	1497	0.32	2512	0	0
184	1480	0.45	1745	17.37	1.3
138	1452	0.64	1594	24.52	3.2
92	1434	0.74	1285	39.15	4.4
46	1297	0.77	1061	49	13.5

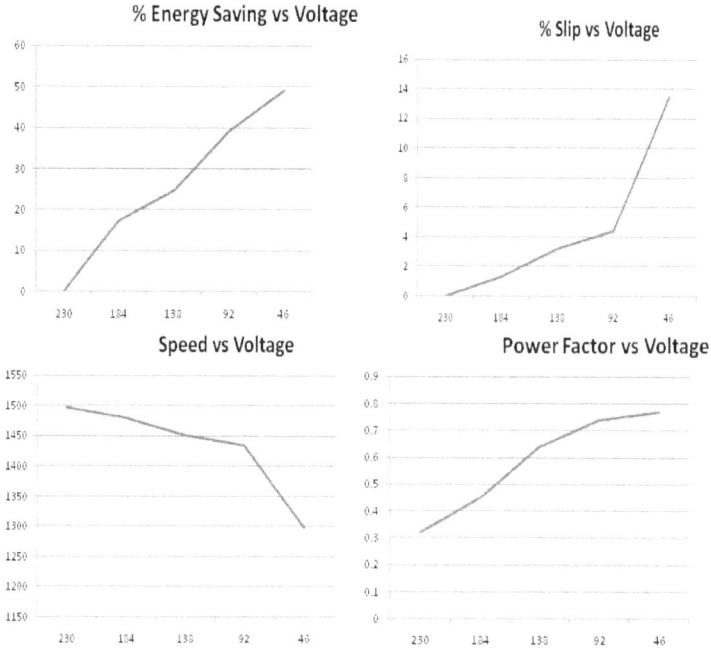

Figure 4.10 Graphs drawn for the values obtained at no load condition.

Above graphs are drawn from the values obtained from the open loop model of the induction motor for energy saving scheme. The graphs indicate that at no load condition the percentage of energy saving, power factor and slip reduces with the increase in the terminal voltage of the motor. With the increase in the voltage, the speed of the motor increases.

Table 4.3 Parameters at partial load condition

Voltage (V)	Losses (W)	% Energy Saving
230	2512	0
184	1992	20.72

Energy saving = (Losses at full voltage - losses at reduced voltage)/ Total losses at full voltage*100   (29)

At no load, Energy saving = (2512-1061)/ 2512* 100= **49%**

At partial load, Energy saving **=** (2512-1992)/ 2512 *100 =**20.72%**

From the Figure 4.10, we infer that with reduction in the terminal voltage from a rated value to 20% of the rated, 49% of energy could be saved and the power factor is observed to be improved from 0.32 to 0.77. The motor speed fell from 1497 to 1297. At partial load condition, 37.02% of the energy is saved by decreasing the terminal voltage of the motor.

Closed Loop Model

The closed loop model of three phase induction machine under no load condition is shown in Figure 4.11. Based on the load torque, the voltage reference is set and the input to the motor is adjusted.

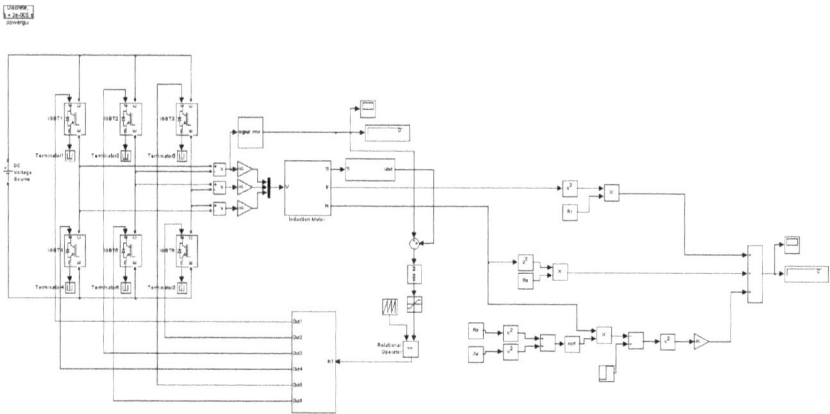

Figure 4.11 Closed loop model of three phase induction machine under no load.

In the closed loop model, various load conditions are assumed. As the load condition varies, the value of the voltage supplied to the motor also varies. During full load condition, full voltage is applied. During partial load condition, only partial voltage is applied. During no load condition, 20% of the voltage is applied. The simulation of the circuit is done for the different loading condition of the motor and the output waveforms for the total losses, power factor and speed of the motor are observed. The energy losses are calculated from the results. By using PWM control technique the applied voltage is controlled. Various graphs are obtained from closed loop model of induction motor. Figure 4.12 shows the graph of load torque vs time. Based on the load variations, the voltage reference is altered.

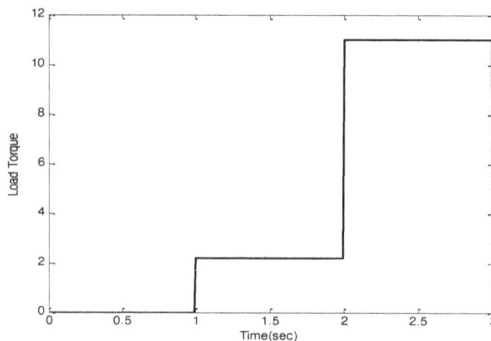

Figure 4.12 Load torque vs. time.

Figure 4.13 shows the graph between voltage reference and time. This voltage wave is obtained with respect to the load variations.

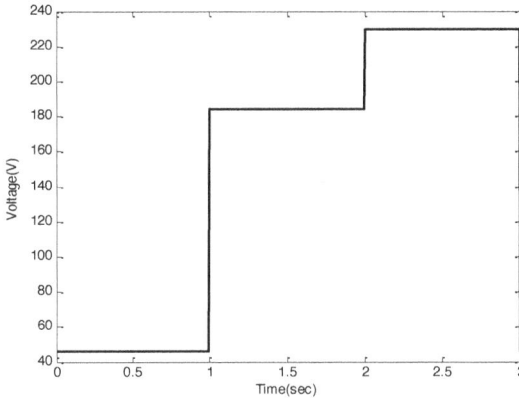

Figure 4.13 Voltage vs. time.

The speed response of the motor and the losses at various load conditions are seen and it is observed that at 20% of the maximum voltage, the speed and losses are reduced. Under no load condition, it is observed that, with the 100% of the rated voltage, the speed of the motor and losses are higher than at 20% of rated voltage and also the power factor is poor with full voltage. Hence, it is noted that the wastage of energy is more when it is operating at rated voltage.

4.3 Conclusion

The energy saving performance of three-phase induction motor drive Space Vector Modulation (SVM) technique is analyzed. The induction motor is modeled using Krause's theory and the same is used in the energy saving calculations. Total losses, power factor, input power and the speed of the induction motor are obtained from the developed model. For various load conditions, the saving in energy has been calculated for different values of the voltage. From the simulation results, it is seen that 49% of energy can be saved under no load condition and 37% at partial load condition. It is also observed that the power factor is improved with reduced voltage operation.

References

1. Tolbert, L. M., and Ozpineci, B. (2005) Simulink implementation of induction machine model – A modular approach. *IEEE*, 728-734.
2. Linga Swamy, R., and Satish Kumar, P. (2008) Speed Control of Space Vectored Modulated Inverter Driven Induction Motor. *Proceedings of the International Multi-conference of Engineers and Computer Scientists, China*, p. 2.
3. Adeel, M. S., Izhar. T., and Saqib, M. A. (2009) An Efficient Implementation of the Space Vector Modulation based Three Phase Induction Motor Drive. *Proceedings the Third IEEE International Conference on Electrical Engineering*, pp. 1-6.
4. Kumar, R., Gupta, R. A., and Surjuse, R. S. (2008) A vector controlled motor drive with neural network based space vector pulse width modulator. *Journal of Theoretical and Applied Information Technology*, **4**, 577-584.
5. Trzynadlowski, M. (1996) An overview of modern PWM techniques for three- phase, voltage-controlled, voltage-source inverters. *Proceedings of the IEEE International Symposium on Industrial Electronics, Poland*, pp. 25-29.
6. Zhang, W.-F., and Yu, Y.-H. (2007) Comparison of three SVPWM strategies. *Journal of Electronics Science and Technology of China*, **5**, 283-287.
7. Shi, K. L., Chan, T. F., Wong, Y. K., and Ho, S. L. (1999) Modeling and simulation of three phase induction motor using Simulink. *International Journal of Electrical Engineering Education*, **36**, 163-172.
8. Jamuna, V., and Rama Reddy, S. (2009) ANN controlled energy saver for induction motor drive. *Journal of Electrical Engineering,* **8**(4), 70-77.
9. Xue, X. D., and Cheng, K. W. E. (2006) An Energy-saving Scheme of Variable Voltage Control for Three-Phase Induction Motor Drive Systems. *2nd International Conference on Power Electronics Systems and Applications,* pp. 241-243.
10. Freeman, J. A., and Skapura, D. M. (2002) *Neural Networks Algorithms, Applications, and Programming Techniques*, Pearson's Education, USA.
11. Asaii, B., Gosden, D. F., and Sathiakumar, S. (1996) Neural Network Applications in Control of Electric Vehicle Induction Machine Drives. *IEE Transactions on Power Electronics and Variable Speed Drives*, **429**, 273-278.

5

Feasibility Study on Fabrication of PLA Dental Crown using 3D Printer

R. Rekha[1], N. Baskar[1], S. Vinoth Kumar[1] and M. V. Suganyadevi[2]

[1]*Department of Mechanical Engineering, Saranathan College of Engineering, Tiruchirapalli, Tamilnadu, India*
[2]*Department of Electrical Engineering, Saranathan College of Engineering, Tiruchirapalli, Tamilnadu, India*

Abstract

There is a continuous demand in the field of dental crown development and methodology in order to replace existing materials. Owing to ceramic's high degree of brittleness, less wear resistance and high cost there is a necessity for a new material to replace ceramics. Several research studies are being carried out to select a proper material and manufacturing process for the dental crown. In this work, survey has been carried out to fulfil the need and demand. Advancement in the technology with composite usage has increased range and scope for use of composites in dentistry. Thereby PLA (poly(lactic acid)) material is found to be the appropriate replacement for the existing material and additive manufacturing is the most feasible and efficient methodology to manufacture the dental crown. PLA is an eco-friendly polymer which is biodegradable, biocompatible and non-toxic in nature. PLA's growth in biomedical sector like Tissue engineering, bone implants and its ease of manufacturing using 3D Printer makes PLA material a best replacement for ceramics in dentistry. As 3D Printing is an additive manufacturing process there will be reduced wastage of material and greater accuracy during manufacturing. 3D printing bestows the user with various customizing options such as varying the density of the product, fabrication of intricate structures with greater accuracy and precision.

5.1 Introduction

In the last few decades there has been a continuous attempt to replace the existing ceramic material in dental crown application, owing to its high cost. Advancement in the technology with composite usage has increased range and scope for use of composites in dentistry. In consideration of strength and money, ceramic material has its own shortcoming than others. **Lu *et al.* [1]** have reported that in ceramic crowns the stiffness and bonding strength of cement agents are reduced due to water aging which in turn reduced the load-bearing capacity of all ceramic restorations after years of service by degrading its bonding strength and stiffness and thereby causing stress redistribution in the restored crown. **Zhang *et al.* [2]** reported that there was appearance of fluorapatite veneering porcelains at the fracture surface. The authors claim that there was more loss due to wear in Monolithic lithium disilicate glass-ceramics with higher mechanical properties. Oscar **Borrero-Lopez *et al.* [3]** stated that prosthetic failure or premature tooth can be caused due to wear damage. It is reported that highest wear rate was observed in lithium disilicate ceramic crowns. **Liu *et al.* [4]** reported that the effect of cementing defects and debonding related defects are more harmful to interfacial stress than the direct effect of debonding. This statement is made after studying the effect of defects and adhesive failure between crown and cement on the stress distribution of all ceramic crowns. **Yin *et al.* [5]** stated that metals are the oldest restorative materials that have been widely used as crowns. They are mechanically strong and durable on the other hand are not aesthetically good looking due to their metallic color. They also have a risk of causing toxic or allergic reactions in the soft and hard tissues of the oral cavity. Lithium disilicate glass ceramics, zirconia and glass-infiltrated alumina and spinel have high-strength but are difficult to machine. Ceramics are widely preferred because of its property of not reacting with enzymes present in mouth but the main disadvantage of the ceramic is that their dental strength is very low, and their load resistance capacity is low. Owing to ceramic's high degree of brittleness, less wear resistance and high cost there is a necessity for a new material to replace ceramics. One of the strong contenders for the described problem is Polylactic Acid (PLA) owing to its robustness and affordability. PLA is a polymer which is eco-friendly due to its biodegradable, biocompatible and non-toxic properties. Medical implants, orthopedic devices, drug delivery system, etc., represent the wide application areas of PLA and its composites. The recent research of the utility of PLA based polymers and their associated materials in Bio-medical applications and the patents obtained are highlighted in this review.

5.2 Literature Review

In order to study the applications of 3D printing and fabrication of dental tooth cap a literature survey has been carried out. A lot of research work has been done in fabrication and evaluation of strength of PLA in various biomedical applications. Various research articles on PLA materials, their properties, their manufacturing using 3D printing and additive manufacturing, their applications on Dental crowns were surveyed in order to have a great knowledge and short coming in the existing works. The following are the summary of the literature survey.

PLA

Poly Lactic acid is a bio-degradable and bio-compatible thermoplastic with good mechanical properties. PLA is harder than Poly tetrafluoroethylene (PTFE) and dimensionally stable at the same time it is inexpensive. PLA is obtained as a natural product by the fermentation of sugar in sugarcane and corn by micro-organisms and is a polymeric form of lactic acid. They are used as an alternative for ceramic based polymeric materials to reduce environmental impact. They are also widely applied as bio-medical materials in areas such as controlled drug delivery system and tissue regeneration. As compared to other traditional biopolymers PLA and its composites have wide biomedical applications like orthopaedics, drug carriers, facial fracture repair, tissue engineering, antimicrobial agents, anti-tumor agents and Urethral stents due to its exceptional properties and its low cost.

D'Agostino *et al.* [6] examined the bone structure and made a PLA copolymer with glycerol cross linker which can be used in bone implants. This bone polymer had good load bearing ability and can help in healing the small fracture. **Rasal *et al.* [7]** created a copolymer of acrylic acid and poly-ethylene glycol with a PLA structure for the usage of backbone. They have very good mechanical structure as well as good toughness in order to withstand a high load in the backbone. **Russias** *et al.* **[8]** observed a reasonable match between the mechanical properties of polymers with ceramic contents ranging between 70 and 85 wt.% and the mechanical properties of human cortical bone. **Wei** *et al.* **[9]** reported that the addition of o-carboxymethyl chitosan (CMC) content increases the tensile and shear moduli of system based on the calculated results of static mechanical properties. This clearly indicates that the rigidity and hardness of blend systems can be improved by the establishment of CMC. With the addition of CMC content, the bulk modulus and Cauchy pressure increased, and it was observed that 8PLA/2CMC has the largest value. Thus, the blend of 8PLA/2CMC possessed the best toughness and fracture resistance. The stronger polar functional groups of CMC and the stronger intermolecular interaction are the reason behind the above phenomena and might have caused the aggregation behavior. **Yu** *et al.* **[10]** found that when the crystallinity is increased from 3.6% to 12% then there is 15% increase in the flexural modulus of FFF (Fused Filament Fabrication) printed parts. It was reported that the crystallization ability of matrix PLA can be improved by the introduction of nucleating agent Talc. Thus, melt crystallization in PLA can still occur regardless of rapid cooling in FFF which may lead to higher degree of crystallinity. **Aravind Raj** *et al.* **[11]** reported that bacterial degradation like Amycolatopsiscoloradensis may cause degradation in PLA. It is said that PLA is very eco-friendly as its manufacturing process involves less emissions and wastages when compared with the conventional processes. Thus, PLA may stand as a competitive material to ABS. **Alam** *et al.* **[12]** stated that PLA nanocomposites along with conductive fillers can be used for scaffold bone replacement and regeneration and other such tissue engineering applications due to their enhanced bioactivity and faster biodegradation response. **Liu** *et al.* **[13]** stated that in upright orientation better printing formability was observed in Virgin PLA when compared with ceramic, copper and aluminium-based PLA. **Rebelo** *et al.* **[14]** showed the direct proportionality of dynamic force plateau and the specific energy absorption with the considered relative density, which causes buckling of interior side cell walls and crushing of PLA honeycomb at the top and bottom layers. The even distribution of the applied blast pressure to the crushable core and its uniform compression can be ensured by the skin plate. Progressive deformation takes place in the crushable core on application of constant low stress which absorbs most of the blast induced energy and thereby protects the load bearing element. **Lee *et al.* [15]** reported that better geometric accuracy was obtained at higher cooling speeds for a specified model in 3D printing. It is also indicated that the cooling air velocity can be considered as an additional control parameter in 3D printing.

Additive Manufacturing

Ezeh et al. [16] have analyzed the strength of Additively Manufactured (AM) Polylacticide (PLA). The experimental results suggested that in PLA, modelling of the effect of the mean stress in fatigue stress can be done by using the maximum stress in the cycle. The stress/strength analysis can be performed effectively by considering this polymer as a homogenous, linear-elastic and isotropic material, since the printing direction appears to have little effect on the overall fatigue behavior of AM PLA. **Papon *et al.* 17]** have done a study on additively manufactured carbon reinforced composites for their fracture toughness. **McGregor *et al.* 18]** have done a study on mechanical properties of hexagonal lattice structures fabricated by additive manufacturing using continuous liquid interface production.

3-D Printing

Material wastage is a serious concern in all form of industries. As 3-D Printing is an additive manufacturing process there will be reduced wastage of material and greater accuracy during manufacturing. Considering this as an important aspect, focus has been given in design stage. Since 3-D printing is taken as the manufacturing technique, the CREAT BOT software which governs the 3-D printing gives the liberty to vary the density. Using this feature as advantage density variation has been done in the tooth design. As load acting on different teeth varies upon their location inside the mouth, research and testing has to be done in order to decide the density requirement. Considering this, density variation is to be implemented to efficiently use the material to its fullest. As dental crown is being manufactured using the process, Fluid Deposition Modelling which is an additive manufacturing process material wastage is reduced to fullest extent. **Martins *et al.* [19]** researched in the field of "Regenerative medicine" which is an emergent field that aims to significantly improve tissue repair or restoration. Three-dimensional (3D) printing is a technique used in Tissue Engineering technology for the synthesis of 3D constructs that play a key role in the physiological environment of cell growth. **Haleem** *et al.* **[20]** proposed the treatment of bone defects with the manufacturing of porous bone scaffold using 3D printing technology and stated that Additive manufacturing technique finds its application in the printing of biomaterials and thus stands as an excellent technology to merge cells and develop their growth factor. In 3D printing, the biomaterial is added layer by layer to form a tissue-like structure. **Anh-VuDo** *et al.* **[21]** analyzed the application of 3-D printing technology in fabricating functional and replicable scaffolds which are capable of promoting tissue engineering. Printing technologies like direct 3D printing, 3D-Bioplotter printing, fused deposition modeling, stereo lithography, selective laser sintering, indirect 3-D printing, and electro spinning can be used for creating scaffolds. **Ngo** *et al.* **[22]** indicated that fused deposition modelling (FDM) is one of the most commonly used 3D printing technique due to its low-cost, high-speed processing and simplicity. It is also reported that PLA, acrylonitrile-butadiene-styrene (ABS) copolymers, polycarbonate (PC), polyamide (PA) and thermosetting powders like polystyrene, polyamides, photopolymer resins are the most commonly used polymers in 3D printing. **Zhao** *et al.* **[23]** Studied on the adhesion behavior of polydopamine coatings on PLA pellets that are manufactured using 3D printing. Research was done for improving the mechanical properties in recycled PLA manufactured using 3D printing. It was claimed that the mechanical properties have improved in the recycled specimen coated with polydopamine. By means of recycling PLA, the manufacturing cost as well as its harmful effects on environment are reduced. **Blok** *et al.* **[24]** investigated the methods to improve the mechanical properties of thermoplastic materials used in Fused Filament Fabrication (FFF) which is one of the 3D printing techniques. Study was done on composite 3D printing feedstock for FFF, in which the carbon fibers are embedded into the thermoplastic matrix for improving its strength and stiffness. **Le Duigou** *et al.* **[25]** and **O'Neil** *et al.* **[26]** have reported that Fused Deposition Modeling technique has low cost, relatively high speed and potential for reinventing the design process. **Aravind Raj** *et al.* **[27]** has studied about the mechanical properties and cost effectiveness of PLA. The article analyses newer 3D Printing Technique and its feasibility to implementation towards biomedical components. The study revealed the PLA is a biodegradable polymer which possesses good strength and is also biodegradable in nature which makes it the most sought after in the production of newer components through

3D Printing. **Rafael *et al.* [28]** has worked in the mechanical characterization of materials produced by 3D printing based on fused filament fabrication. The materials chosen was polylactic acid (PLA) and a PLA reinforced with short carbon fibers in a weight fraction of 15% (PLACE). The literature reveals that very limited work has been carried out in fabrication and strength analysis of PLA dental crown. Moreover PLA has superior properties which make them suitable for biomedical applications. **Xu *et al.* [29]** reported that the shear-thinning behavior can be improved by incorporating various nanoscale additives into the hydrogel. This makes it more suitable for extrusion-based 3D printing thereby improving the printability of these high-strength hydrogels. These interactions also help in maintaining the desired 3D shapes and in increasing the mechanical properties of the hydrogels.

5.3 Research Gap

- o The major problem faced in case of ceramic is the high degree of brittleness, less wear resistance and high cost. From the market survey it has been found that the durability of the ceramics is very less compared to all other material. On usage they lose their color in case of appearance.
- o All the ceramics cannot be placed in all positions, to be specific in rear molar. These ceramics cannot be used in that position because it requires high load bearing capacity during the chewing process.
- o As the ceramics are brittle in nature and have high hardness, the ceramic teeth tend to get damaged during impact.
- o Repairing of the ceramic tooth is not possible and it has to be replaced as such rather than remodeling.
- o In metal ceramic crowns there may be occurrence of failures such as cohesive, adhesive and mixture failure.

5.4 Conclusion

From the above research it can be concluded that PLA is an appropriate replacement for the existing material in owing to its greater strength and less cost. The combined effects of PLA's strength, resilience and pliability makes its scientific importance inevitable in the field of dentistry and medicine.3-D printing is the best methodology among the existing ones because of its accuracy and its feature which enables to vary the density of the material. Fluid deposition modelling used in 3D printing is an additive manufacturing process and helps in reduction of wastage to a greater extent. The design and fabrication using 3D printer is applicable and user friendly for all the dentists. So, it is feasible to fabricate the dental tooth with a greater accuracy and strength using 3D printing as the method to manufacture and PLA as a composite raw material.

5.5 References

1. Lu, C., Wang, R., Mao, S., Arola, D., and Zhang, D. (2013) Reduction of load-bearing capacity of all-ceramic crowns due to cement aging. *Journal of the Mechanical Behavior of Biomedical Materials*, **17**, 56–65.
2. Zhang, Z., Yi, Y., Wang, X., Guo, J., Li, D., He, L., and Zhang, S. (2017) A comparative study of progressive wear of four dental monolithic, veneered glass-ceramics. *Journal of the Mechanical Behavior of Biomedical Materials*, **74**, 111–117.
3. Borrero-Lopez, O., Guiberteau, F., Zhang, Y., and Lawn, B. R. (2019) Wear of ceramic-based dental materials. *Journal of the Mechanical Behavior of Biomedical Materials*, **92**, 144-151.
4. Liu, Y., Xu, Y., Su, B., Arola, D., and Zhang, D. (2018) The effect of adhesive failure and defects on the stress distribution in all-ceramic crowns. *Journal of Dentistry*, **75**, 74–83.
5. Yin, L., Song, X. F., Song, Y. L., Huang, T., and Li, J. (2006) An overview of in vitro abrasive finishing & CAD/CAM of bioceramics in restorative dentistry. *International Journal of Machine Tools and Manufacture*, **46**(9), 1013-1026.
6. D' Agostino, J. A., and Watterson, A. (2012) Implantable polymer for bone and vascular lesions, WO2012054742.
7. Rasal, R. M., and Hirt, D. E. (2010) Copolymer including polylactic acid, acrylic acid and polyethylene glycol and processes for making the same, WO2010056421.
8. Russias, J., Saiz, E., Nalla, R. K., Gryn, K., Ritchie, R. O., and Tomsia, A. P. (2005) Fabrication and mechanical properties of PLA/HA composites: A study of in vitro degradation. *Materials Science and Engineering: C*, **26**(8), 1289-1295.
9. Wei, Q., Cai, X., Guo, Y, Wang, G., Guo, Y., Lei, M., Song, Y., Yingfeng, Z., and Wang, Y. (2019) Atomic-scale and experimental investigation on the micro-structures and mechanical properties of PLA blending with CMC for additive manufacturing. *Materials & Design*, **183**(5). 108158.

10. Yu, W., Wang, X., Ferraris, E., and Zhang, J. (2019) Melt crystallization of PLA/Talc in fused filament fabrication. *Materials & Design*, **182**, 108013.
11. Aravind Raj, S., Muthukumaran, E., and Jayakrishna, K. (2018) A case study of 3D printed PLA and its mechanical properties. *Materials Today: Proceedings*, **5**(5), 11219-11226.
12. Alam, F., Varadarajan, K. M., and Kumar, S. (2019) 3D printed polylactic acid nanocomposite scaffolds for tissue engineering applications. *Polymer Testing*, **81**, 106203.
13. Liu, Z., Lei, Q., and Xing. S. (2019) Mechanical characteristics of wood, ceramic, metal and carbon fiber-based PLA composites fabricated by FDM. *Journal of Materials Research and Technology*, **8**(5), 3741-375.
14. Rebelo, H. B., Lecompte, D., Cismasiu, C., Jonet, A., Belkassem, B., and Maazoun, A. (2019) Experimental and numerical investigation on 3D printed PLA sacrificial honeycomb cladding. *International Journal of Impact Engineering*, **131**, 162-173.
15. Lee, C.-Y., and Liu, C.-Y. (2019) The influence of forced-air cooling on a 3D printed PLA part manufactured by fused filament fabrication. *Additive Manufacturing*, **25**, 196-203.
16. Ezeh, O. H., and Susmel, L. (2018) On the fatigue strength of 3D-printed polylactide (PLA). *Procedia Structural Integrity*, **9**, 29-36.
17. Papon, E. A., and Haque, A. (2019) Fracture toughness of additively manufactured carbon fiber reinforced composites. *Additive Manufacturing*, **26**, 41-52.
18. McGregor, D. J., Tawfick, S., and King, W. P. (2019) Mechanical properties of hexagonal lattice structures fabricated using continuous liquid interface production additive manufacturing. *Additive Manufacturing*, **25**, 10-18.
19. Martins, J. P., Ferreira, M. P. A., Ezazi, N. Z., Hirvonen, J. T., Santos, H. A., Thrivikraman, G., França, C. M., Athirasala, A., Tahayeri, A., amd Bertassoni, L.E. (2018) 3D printing: prospects and challenges. In: *Nanotechnologies in Preventive and Regenerative Medicine*, Uskokovic, V., and Uskokovic, D. P. (eds.), Elsevier, USA, pp. 299-379.
20. Javaid, M., and Haleem, A. (2019) 3D printed tissue and organ using additive manufacturing: An overview. *Clinical Epidemiology and Global Health*, 8, 586-594.
21. Do, A.-V., Smith, R., Acri, T. M., Geary, S. M., and Salem, A. K. (2018) 3D printing technologies for 3D scaffold engineering. In: *Functional 3D Tissue Engineering Scaffolds: Materials, Technologies and Applications*, Deng, Y., and Kuiper, J., Elsevier, USA, pp. 203-234.
22. Ngo, T. D., Kashani, A., Imbalzano, G., Nguyen, K. T. Q., and Hui, D. (2018) Additive manufacturing (3D printing): A review of materials, methods, applications and challenges. *Composites Part B: Engineering*, **143**(15), 172-196.

23. Zhao, X. G., Hwang, K.-J., Lee, D., Kim, T., and Kim. M. (2018) 'Enhanced mechanical properties of self polymerisedpolydopamine-coated recycled PLA filament used in 3D printing. *Applied Surface Science*, **441**, 381-387.

24. Blok, L. G., Longana, M. L., Yu, H., and Woods, B. K. S. (2018) An investigation into 3D printing of fibre reinforced thermoplastic composites. *Additive Manufacturing*, **22**, 176–186.

25. Le Duigou, A., Barbe, A., Guillou, E., and Castro, M. (2019) 3D printing of continuous flax fibre reinforced biocomposites for structural applications. *Materials & Design*, 180, 107884.

26. O'Neil, G. D., Ahmed, S., Halloran, K., Janusz, J. N., Rodríguez, A., Terrero Rodríguez, I. M. (2018) Single-step fabrication of electrochemical flow cells utilizing multi-material 3D printing. *Electrochemistry Communications,* **99**, 56-60.

27. Aravind Raj, S., Muthukumaran, E., and Jayakrishna, K. (2018) A case study of 3D printed PLA and its mechanical properties. *Materials Today: Proceedings*, **5**(5), 11219-11226.

28. Ferreira, R. T. L., Amatte, I. C., Dutra, T. A., and Bürger, D. (2017) Experimental characterization and micrograph of 3D printed PLA and PLA reinforced with short carbon fibers. *Composites, Part B: Engineering*, **124**, 88-100.

29. Xu, C., Dai, G., and Hong, Y. (2019) Recent advances in high-strength and elastic hydrogels for 3D printing in biomedical applications. *Acta Biomaterialia*, **95**(1)5 0-59.

6

Transformer Monitoring System with an Alternate Power Source using IOT

M. Prakadeesh[1], C. Vennila[2] and Vijay R.[3]

[1]Crayon Data Pvt. Ltd., India
[2]Department of ECE, Saranathan College of Engineering, Trichy, India
[3]Department of EEE, Saranathan College of Engineering, Trichy, India

Abstract

Transformers are the most important component of any power system. Any damage to transformers has a negative impact on a power system's equilibrium. Overloading and poor cooling are the main causes of damage. The major goal of the is to use IOT technology to monitor the health of the distribution transformer in real time. Temperature, voltage, and current of a transformer are all monitored, analyzed, and recorded by servers. We use three sensors connected to an Arduino for this. The recorded data can be sent over Wi-Fi module and viewed via HTTP protocol from anywhere in the world utilizing IOT technology. This aids in identifying without relying on humans. This aids in identifying and resolving an issue before it becomes a failure, without the need for human intervention.

6.1 Introduction

The low voltage users are supplied directly by the distribution transformer. As a result, the transformer's operational condition is critical in the distribution network. For a long life, transformers must be operated in rated condition. This is not feasible for the duration of the working day. Overloading and inadequate cooling of transformers can result in unexpected transformer failure, disrupting energy delivery to a large number of consumers. The manual checkup of voltage, ambient temperature, load current, and other parameters is more difficult because incidental parameters cannot be accessed [1].

Sensors and actuators are used to interface between the physical and digital worlds in IoT. The physical parameters or the respective environment are sensed using a sensor or a network of sensors. With the help of various network devices, these processed sensor outputs are subsequently sent to the main server or cloud. The data can be accessible via the internet from any location on the planet. The primary goal of IoT technology is to monitor and control. As a result, IoT-based monitoring is preferable than manual monitoring. The system monitors transformer parameters such as voltage, current, and temperature in real time. This will aid in the detection of flaws before a major failure happens.

6.2 Faults in Transformers

Overload, over/under voltage, temperature rise, oil level fault, and other severe defects can occur in a transformer, such as

- Overload / Overcurrent: Overload / Overcurrent is the flow of fault current occurring in the power system through the transformer. This condition last for a short duration of about or less than 2 seconds as protection relays isolate the power system.
- Temperature Rise: Transformers are typically intended to operate for 24 hours at a temperature of 300 degrees Celsius. Overvoltage and current produce an increase in oil temperature, which leads to transformer winding insulation breakdown.

6.3 Related Works

One of the most significant pieces of equipment in a power network is distribution transformers. Data capture and condition monitoring are major issues in power electric systems because of the huge number of transformers scattered across a big area. The major goal of this system is to use a GSM modem to monitor and manage distribution transformers. Transformers have been damaged as a result of the oil spill. Different characteristics and environmental variables influence oil deterioration. We're currently focusing on the temperature of the transformer and the viscosity of the oil in this system.

The AVR microcontroller is used to monitor and manage the temperature and viscosity in this system [1-3].

6.4 Hardware Requirement

The system consists of various components:

a) Current Transformer

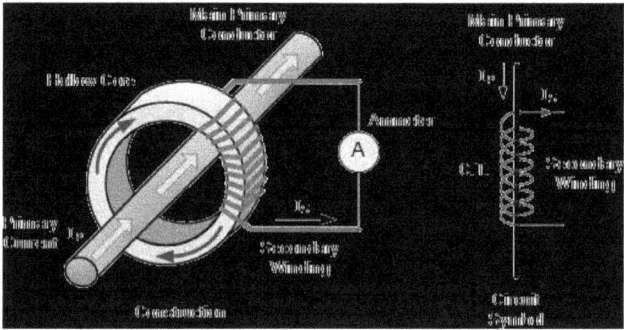

Figure 6.1 Current transformer.

The current transformer (C.T.) produces an alternating current proportional to the current measured in its main winding in its secondary winding. Current transformers reduce high-voltage currents to a safe level, allowing a standard ammeter to safely detect the actual electrical current flowing in an AC transmission line. A basic current transformer differs from a normal voltage transformer in that it functions on a different premise.

b) Potential Transformer

Figure 6.2 Potential transformer.

The most commonly utilized devices are PTs or VTs. Traditional transformers with two or three windings are used in these devices (one primary with one or two secondary). They have an iron core, and the primary and secondary are magnetically coupled. The high side winding has more copper turns than the secondary, and any voltage applied to the primary winding is reflected in direct proportion to the turns ratio or PT ratio on the secondary windings. Voltage transformers are utilized for line and circuit protection and are also known as voltage transformers.

c) Level Sensor

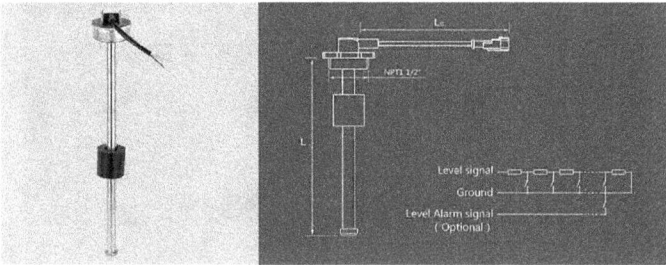

Figure 6.3 Level sensor.

Level sensors are used to measure and control the amount of a free-flowing fluid in a contained space. Although level sensors are typically used to monitor liquids, they can also be used to detect solids such as powdered compounds. Industrial level sensors are widely utilized. Liquid level sensors are used in automobiles to monitor a range of liquids, including petrol, oil, and, on rare occasions, specialty fluids like power steering fluid. A switch at the top and bottom of the container allows for the sensing of minimum and maximum liquid levels.

d) Temperature Sensor

Figure 6.4 Temperature sensor.

The LM35 is an integrated circuit temperature sensor that produces a proportional output based on the temperature (in °C). The sensor circuitry is not subject to oxidation or other processes because it is sealed. Using the LM35, temperature can be sensed more precisely than with a thermistor. It also has a poor self-heating characteristic, with a temperature rise in still air of less than 0.1 °C. The operating temperature range is -55°C to 150°C. With a scale factor of 0.01V/°C, the output voltage changes by 10mV for every °C change in ambient temperature.

e) Microcontroller

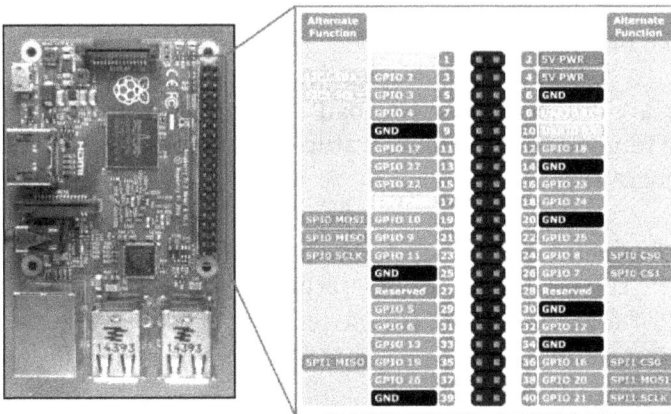

Figure 6.5 Raspberry Pi with Pin diagram.

The system uses raspberry pi and has a microcontroller. The raspberry pi uses raspberry Jessie as operating system. The main advantage of using raspberry pi as microcontroller, the raspberry has faster computation time than the other microcontroller and for IoT application its performance is good with less complications since the requirement of external hardware devices are reduced. The raspberry pi has an inbuilt function of wifi/Ethernet connectivity four USB port and inbuilt Bluetooth function. This specification makes microcontroller for a quick access of internet.

6.4 Block Diagram

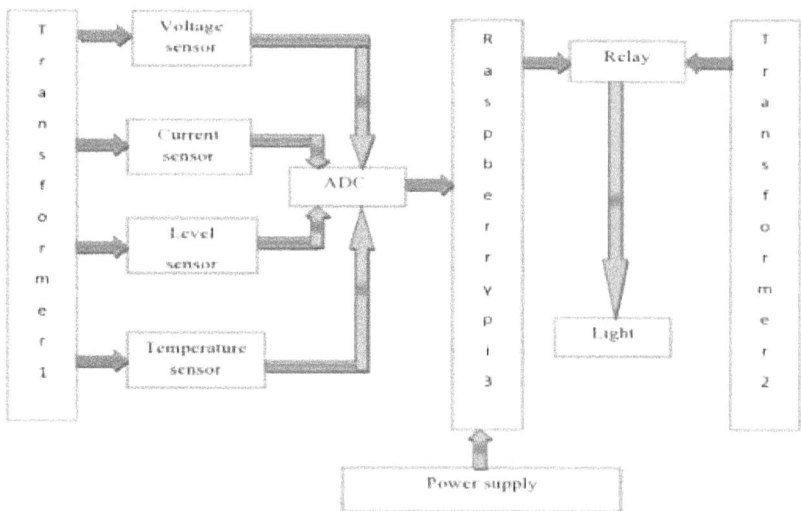

Figure 6.6 Block diagram.

Sensors are utilized to sense load current, ambient temperature, winding temperature, oil temperature, and oil level, according to the block diagram.

- To determine the top oil temperature, use the PT100
- PT100 for ambient temperature
- C.T for load current (single phase)
- Determination of voltage at the bushing's measurement tap (three phase)

All the sensor values will be sent to ADC where it coverts analog values to digital values and feed it on Raspberry pi 3 which monitors the values and update it in the web page. The system sends an alert message to the person in charge if any abnormalities occur in the parameter values. If the transformer 1 fails, then relay switch changes the power supply from Transformer 1 to the Transformer 2. The changes are updated on the web through IoT technology.

6.5x Hardware Implementation

o   The rapsberry pi 3 kit is loaded with the rapsberry jessie OS and the monitor is connected to the kit.The power supply is given through the usb port. Keyboard and mouse are connected to the USB port of the kit.

o   Transformer and various sensors are connected to the ADC. The values are converted to digital and given to the raspberry pi 3 kit. The conditions are monitored and if any abnormalities occurred respective actions are performed.

o   After hardware connection the code is executed, and the various parameters are monitored, and the measured values are displayed. If any abnormalities occurred, the alternate power source is given to the device (bulb) from the nearby transformer. The switching is done automatically, and it is also controlled through IoT.

6.6 Result of Implementation

If any abnormalities occurred in the distribution transformer, then a text message is sent to the authority person for action before any catastrophic failure. The text message is sent through way2 SMS application. A user (mobile number) can send a maximum of 30 text per day. A sample code is written and executed through way2 SMS application, as shown in Figure 6.8.

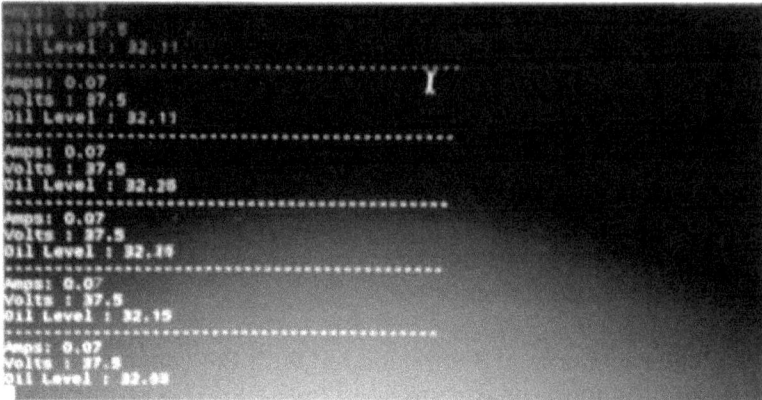

Figure 6.8 The monitored values of current, voltage and oil level of the transformer.

6.7 Future Development

An android application can be developed to monitor the changes in the various parameters values and the switching process can also be done through the application from anywhere through IoT. The monitored values can be stored in the cloud computing since the data can be retrieved when required.

6.8 Present System

In transformer monitoring system, the present technology is SCADA. SCADA is used monitor the various parameters of the transformer and send the results to the sub stations. The solutions can be made by analyzing the results. Our system provides an immediate solution by using an alternate transformer through IOT.

6.9 Conclusion

Transformers are critical components of the power distribution system. As a result, transformer monitoring, and protection are critical. This system uses IoT to bring a new and enhanced technique of monitoring transformer health parameters. Transformer health metrics like as voltage, temperature, and current are collected by the sensors in the system. These data are sent to ThingSpeak, an IoT platform. The HTTP protocol can be used to send and receive data.

The system, thus, allows for real-time data collection, storage and monitoring of transformer health indicators.

References

1. Vadirajacharya, K., Kharche, A., Kulakami, H., and Landage, V. (2012) Transformer health condition monitoring through GSM technology. *International Journal of Scientific & Engineering Research*, **3**(12).
2. Jayakumar, J., Queen, J. H. J., James, T., Hemalatha, G., and Lonappan, N. (2013) Distribution transformer monitoring through GPRS. *International Journal of Scientific & Engineering Research*, **4**(6).
3. Pandey, R. K., and Kumar, D. (2013) Distributed transformer monitoring system based on Zigbee technology. *International Journal of Engineering Trends and Technology*, **4**(5).
4. Zou, L. (2013) Real-time monitoring system for transformer based on GSM. In: *International Conference on Information Computing and Applications: Information Computing and Applications*, Yang, Y., Ma, M., and Liu, B. (eds.), Springer, Germany, pp. 314-323.
5. Gowri, K., Thangam, T., and Rajasaranya (2013) Transformer -An integrated system for monitoring of power transformers. *International Journal of Computer Science and Mobile Computing*, **2**, 105-110.
6. Sujatha, M. S., and Vijay Kumar, M. (2011) On-line monitoring and analysis of faults in transmission and distribution lines using GSM technique. *Journal of Theoretical and Applied Information Technology*, **33**(2).
7. Jeya Padmini, J., and Kashwan, K. R. (2015) Effective Power Utilization and Conservation in Smart Homes Using loT. *2015 International Conference on Computation of Power, Energy, Information and Communication*.
8. Ashraf, Q. M., Hadi Habaebi, M., Sinniah, G. R., Ahmed, M. M., and Khan, S. (2014) Autonomic Protocol and Architecture for Devices in Internet of Things. *2014 IEEE Innovative Smart Grid Technologies*.

7

Mathematical Modelling and Design Implementation of PV Module in MATLAB

A. R. Danila Shirly

Department of EEE, Loyola-ICAM College of Engineering & Technology, Chennai, India

Abstract

This chapter presents the detailed solar panel mathematical modelling and implementation of the same using MATLAB/Simulink. Solar energy is considered as the most preferred non-exhaustible energy owing to its numerous advantages when compared to that of other non-conventional sources. Hence its proper modelling and design knowledge has to be analyzed in detail for effective implementation in real time world. With advent of Simulink technologies, the PV module implementation can be done, and its effect to various irradiation and temperature can be analyzed. The designed PV module can be connected with an appropriate dc-dc converter for impedance matching between the source and the load so that maximum power gets transferred. This chapter helps the reader to understand the design modelling and implementation of PV module using MATLAB/Simulink.

7.1 Introduction

PV

PV stands for photovoltaic which is used to harness energy from sun's irradiation. PV cell converts the solar irradiation that falls on the cell into electricity, without leading to air pollution or noise pollution when compared to the conventional ways of producing electricity, which makes PV cell popular. Apart from the advantages mentioned above it produces electricity even without any moving parts or rotating devices which makes the construction of a PV module robust hence it is popular even among the renewable sources.

PV Cell

A photovoltaic cell or PV cell is a semiconductor device that uses the photoelectric effect to convert light energy into electrical energy. The photovoltaic or photoelectric effect is nothing but conversion of photons that is present in sun's irradiation into flow of electrons (current) in the semiconductor device. In more detail, when the energy of a photon of light exceeds the energy of the band gap, the electron is emitted, and the flow of electrons produces current. PV cell is the fundamental building block of a PV module. PV cells are made of semiconductor devices which have photovoltaic effect namely Silicon and Germanium. Silicon is predominantly used as PV cell over Germanium as it has best photovoltaic effect characteristics. When irradiation strikes the surface of solar cell, the electrons and holes are created with the aid of using the infringement of covalent bond present in the atom of silicon material and as an end result the ability is advanced through developing positive and negative terminals. These terminals when related in series with the load, the current flows via them hence providing power to the load.

PV Module

A PV module includes many PV cells linked in parallel to increase current and in series to supply a better voltage. Usually some of PV cells are organized in series and parallel to fulfil the power requirements. A single cell generates a very low voltage (approx. 0.4) so that more than one photovoltaic cell is connected. Normally a PV module comprises of either 36 or 76 cells depending on the manufacturer. The forward-facing surface of the PV module is maintained transparent and made with low iron and glass content which is entirely encapsulated in a frame. The PV module efficiency is usually degraded to a little extent when compared to PV cell due to the presence of glass and encapsulated frame.

PV Array

A PV array includes some of individual PV modules or panels which have been wired collectively in a sequence and/or parallel to supply the voltage and ampere a specific device requirement. Array can either cover acres of PV modules together or an interconnection of a pair of PV module. The generated power from PV module is not sufficient to supply the grid or trade the power generated, so the PV modules are connected in sequence (series) & in non-sequence (parallel) connection appropriately to form the PV array and supply the grid with power. To achieve more current PV modules are inter-linked in parallel manner and to achieve desired voltage the PV modules are connected in series fashion.

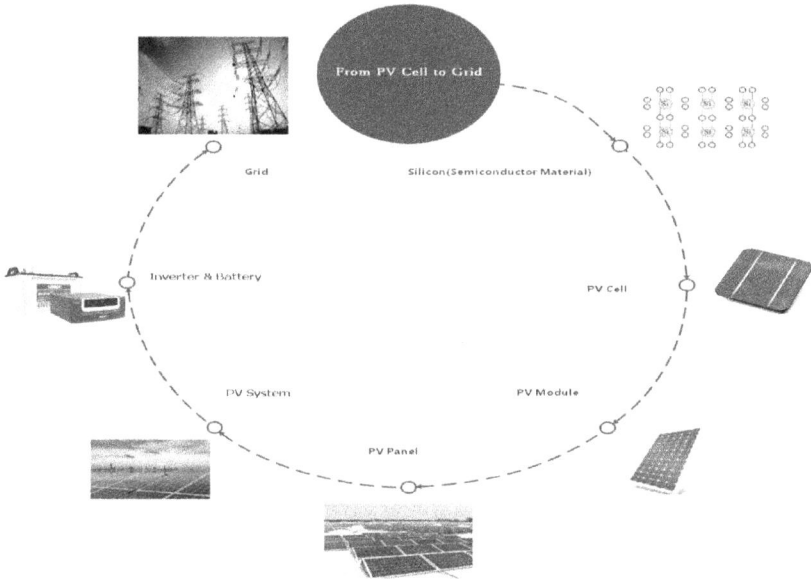

Figure 7.1 From PV cell to grid.

7.2 Fundamental Parameters of PV Cell

The solar cell's fundamental parameters are as follows:

- Efficiency
- Maximum power
- Short circuit current
- Fill factor
- Open circuit voltage

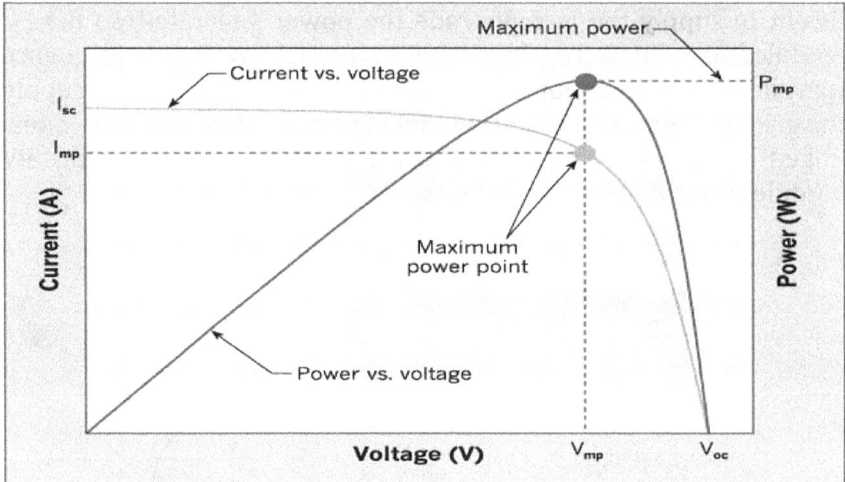

Figure 7.2 I-V and P-V curve of solar cell.

Short Circuit Current (I_{sc})

Short circuit current is the current that flows through a PV cell when the potential across it is zero (i.e., when the solar cell is short circuited).

The following are the factors that influence short-circuit current:

- The solar cell's surface area.
- The quantity of photons
- The incident light's spectrum.
- Optical characteristics.
- The likelihood of photons colliding on the surface of a solar cell, which is primarily determined by two criteria: surface passivation and the minority charges carrier lifetime present in the base.

Open Circuit Voltage (V_{oc})

 Open circuit voltage V_{oc} is the maximum voltage present across the PV cell at zero current. This voltage is analogous to the total amount of forward bias that occurs on the PV cell as a result of the biasing of the semiconductor junction as well as the generation of current.

Maximum Power (P$_{mp}$)

Maximum power is the product of maximum current and maximum voltage as shown in the Figure 7.2.

Efficiency

Efficiency of the PV cell is given importance owing to the expenditure of the PV cell. It is likely to have a PV cell with high efficiency as it is costly, but the fact is that the PV cell's efficiency is very low. The efficiency is the parameter used to compare the performance of one type of solar cell to another. The efficiency of a solar cell is defined as the ratio of the output energy extracted from the cell to the input energy incident on the cell from the sun. This parameter is much required apart from analyzing the PV cell performance it depends on certain crucial factors like spectrum, the solar irradiation intensity and the temperature of the PV cell. As a result, the efficiency must be measured under appropriate atmospheric conditions.

$$\eta = \frac{V_{OC} I_{SC} FF}{P_{in}}$$

Fill Factor

The "fill factor," commonly abbreviated as "FF," is a critical parameter that, when combined with V_{oc} and I_{sc}, allows us to calculate the maximum power obtained from a solar cell. Technically, the FF is defined as the ratio of the maximum power generated by a PV cell to the product of V_{oc} and I_{sc}.

$$FF = \frac{V_{MP} I_{MP}}{V_{OC} I_{SC}}$$

7.3 Basic Equivalent Circuit of Solar Cell

Figure 7.3 depicts the equivalent circuit of a perfect PV cell. It is equipped with a single pn junction diode that is connected in parallel with the generated light source. The PV step-by-step procedure for implementing the mathematical model of a PV cell in MATLAB is provided, along with the equation and the corresponding Simulink/MATLAB model of each equation that defines the PV array.

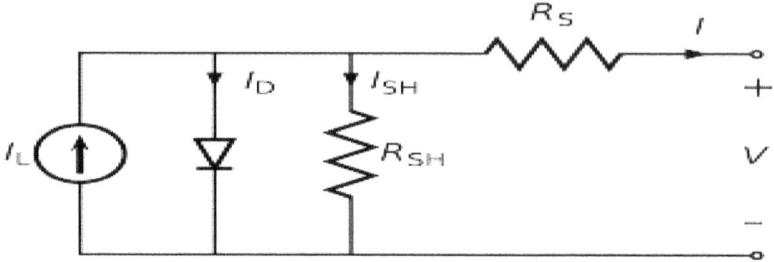

Figure 7.3 Equivalent circuit of solar cell.

Thermal Voltage equation

$$V_t = \frac{T_{op} \times K}{q}$$
(1)

where,
T_{op} is operating temperature [K]
q is electron charge (q=1.602e-19[C])
K is Boltzmann constant (K=1.3806e-23[J/K])

Eq. 1 is depicted in Simulink/MATLAB as shown in Figure 7.4.

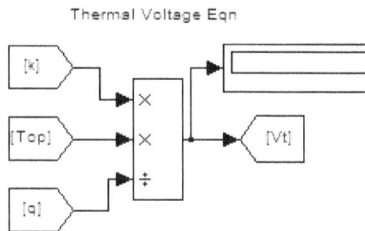

Figure 7.4 Thermal Voltage model.

Reverse saturation current,

$$I_{rs} = \frac{I_{sc}}{(\exp((V_{oc} \times q)/(K \times C \times T_{op} \times n)) - 1)} \qquad (2)$$

where,
V_{oc} is Open Circuit voltage [V]
I_{sc} is Short Circuit current [A]
n is ideality factor (1.3)
C is number of cells

Eq. 2 is depicted in Simulink/MATLAB as shown in Figure 7.5.

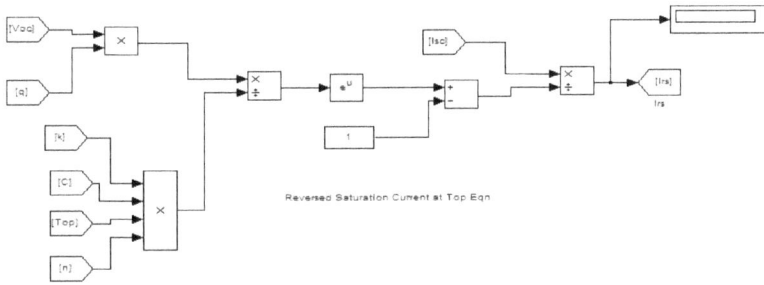

Figure 7.5 Reverse saturation current model.

Saturation current

$$I_s = I_{rs} \left(\frac{T_{op}}{T_{ref}}\right)^3 \exp\left\{\frac{q^2 \times E_g}{K \times n} \times \left(\frac{1}{T_{op}}\right) - \left(\frac{1}{T_{ref}}\right)\right\} \qquad (3)$$

where,
T_{ref} is reference temperature [25+273] [K]

Eq. 3 is depicted in Simulink/MATLAB as shown in Figure 7.6.

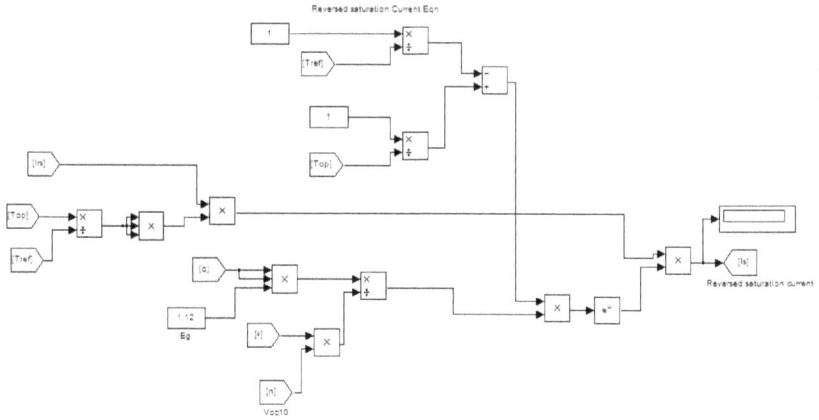

Figure 7.6 Saturation current model.

Diode current

$$I_d = I_s \times N_p \times \left[\exp\left(\dfrac{\dfrac{V}{N_s} + \left[\dfrac{I \times R_s}{N_s}\right]}{n \times V_t \times C} \right) - 1 \right]$$
(4)

where,
N_s is number of modules in series
N_p is number of modules in parallel

Eq. 4 is depicted in Simulink/MATLAB as shown in Figure 7.7.

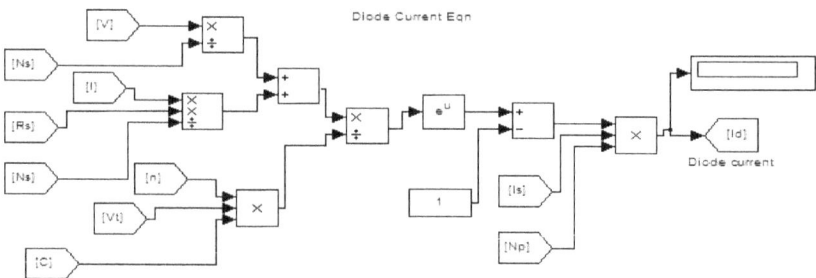

Figure 7.7 Diode current model.

Shunt current,

$$I_{sh} = \left[\left(\frac{V + (I \times R_s)}{R_p} \right) \right]$$

(5)

where,
R_s is Series resistance [Ω]
R_p is Parallel resistance [Ω]

Eq. 5 is depicted in Simulink/MATLAB as shown in Figure 7.8.

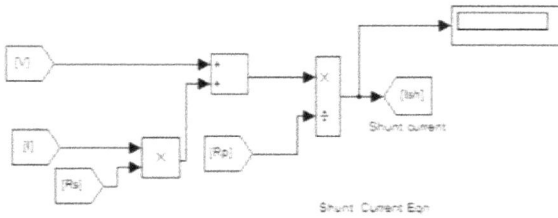

Figure 7.8 Shunt current model.

Phase current

$$I_{ph} = \left\{ \left[\left(T_{op} - T_{ref} \right) \times K_I \right] + I_{sc} \right) \times I_{rr} \right\}$$

(6)

where,
K_i is current temperature coefficient [A/K]
K_v is voltage temperature coefficient [V/K]

Eq. 6 is depicted in Simulink/MATLAB as shown in Figure 8.9.

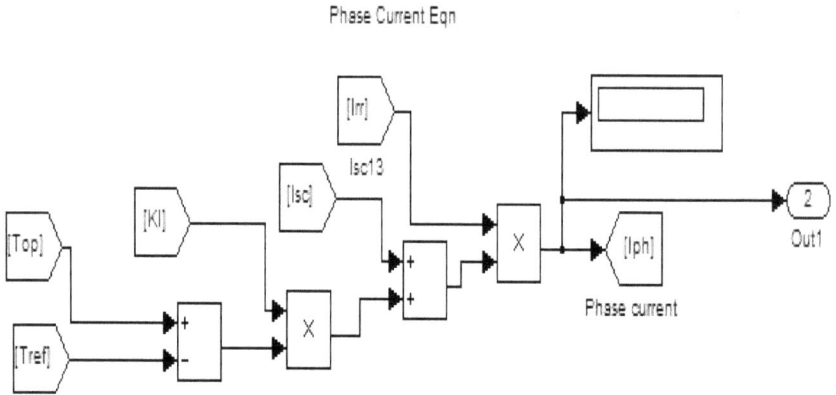

Figure 7.9 Phase current model.

Load current

$$I = \left\{ \left| I_{ph} \times N_p \right| - I_{sh} - I_d \right\} \tag{7}$$

Eq. 7 is depicted in Simulink/MATLAB as shown in Figure 7.10.

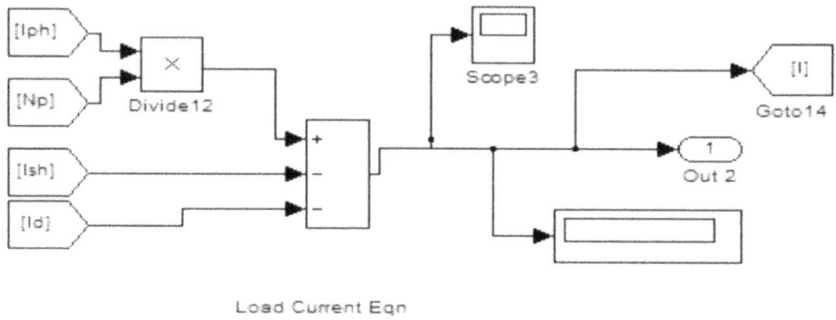

Figure 7.10 Load current model.

Complete PV Block

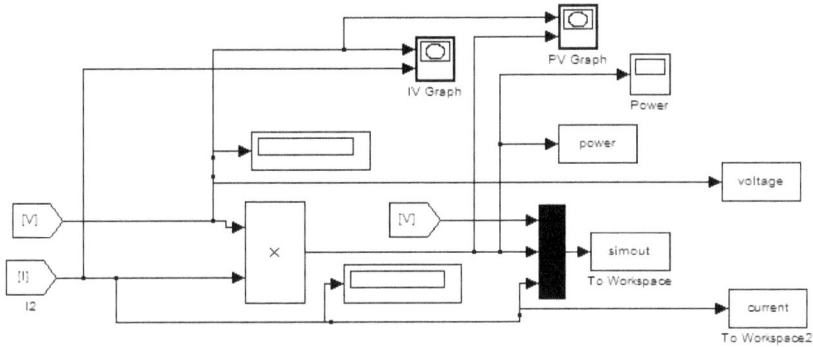

Figure 7.11 Complete PV model.

Table 1 Panel specifications

Voltage at Maximum Power (Vmpp)	18.35[V]
Current at Maximum Power (Impp)	1.674 [A]
Open Circuit Voltage (Voc)	22.2 [V]
Short Circuit Current (Isc)	1.73 [A]
Output Power (Po)	30.0 [W]

Thus, by giving the panel specifications in the modelled PV module IV and PV plot of the solar can be easily plotted using MATLAB.

7.4 The Influence of Variations in Solar Irradiation

The solar irradiation has a significant influence on the P-V and I-V curves of the PV cell. The light incident on the panel is the result of environmental changes that occur in the atmosphere, which is constantly changing, so in order to extract the maximum irradiation certain control techniques are used so that irradiation can be tracked, and the power demand required by the load can be met during the fluctuating weather conditions. Solar irradiation is directly proportional to open circuit voltage; as irradiation increases, so does the open circuit voltage. The reason behind this theory is that when the PV cell is exposed to more sunlight the electrons in the PV cell are energized with high excitation level which makes the electrons go mobile with high speed thereby increasing the generation of power. The variation of P-V and I-V plot with respect to change in solar irradiation is shown in FIGURE 12 & FIGURE 13 respectively.

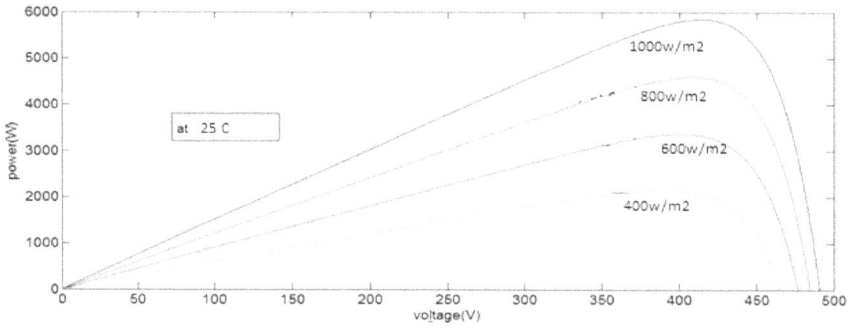

Figure 7.12 Variation of P-V curve with solar irradiation.

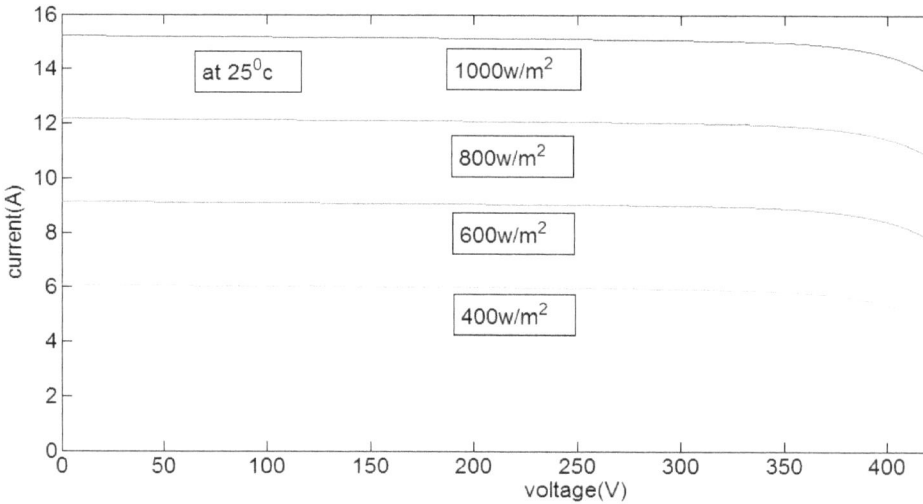

Figure 7.13 Variation of I-V curve with solar irradiation.

7.5 The Influence of Temperature Variation

Temperature effect on PV cell is inversely proportional to its output generated from the PV cell, which means as the temperature of the PV cell increases the energy extracted from it gets reduced which is contrary to the effect produced by solar irradiation. The open circuit voltage of the PV cell decreases as the temperature rises. This is due to an increase in the band gap level, which requires more energy by the electrons to cross this barrier. As a result, the PV cell's performance degrades as the temperature rises, and the efficiency of the PV cell decreases. The temperature effect on the P-V and I-V curve of the solar cell is shown in Figures 7.14 and 7.15 respectively.

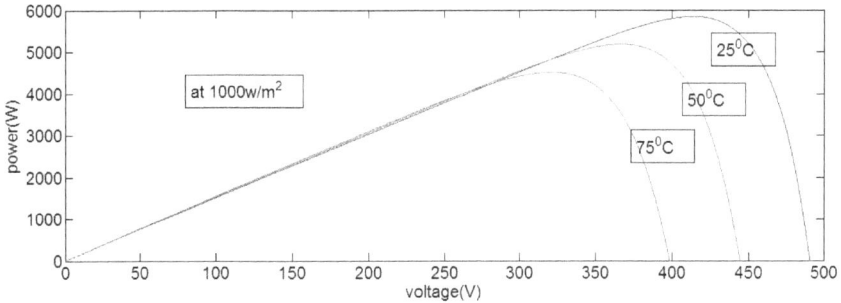

Figure 7.14 Variation of P-V curve with temperature.

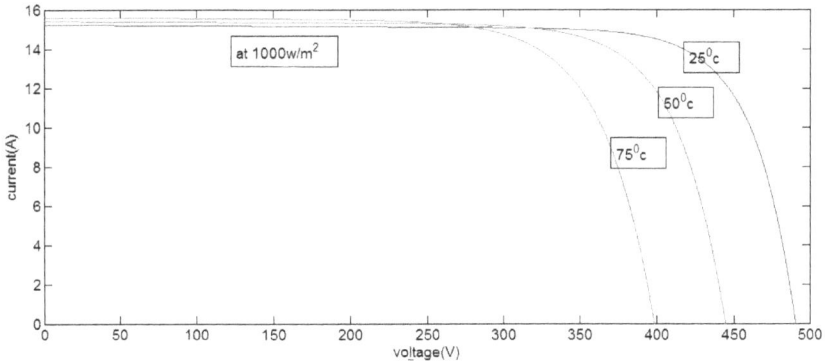

Figure 7.15 Variation of I-V curve with temperature.

7.6 Conclusions

The chapter deals with a thorough mathematical modelling of PV module and step by step implementation of PV using MATLAB. The solar panel's performance is plotted for various insolation/irradiation and various temperatures. This article specifies the fundamental technologies used in PV for better understanding of the characteristic of panel. The I-V and P-V curve plot was done for standard test condition and also for various irradiation and temperature settings. The chapter further opens a new area of research work where incorporation of any advanced dc-dc converter for impedance matching and proper MPPT control or any other metaheuristic approach can be implemented, and the maximum power can be tracked from the panel thereby using the module effectively for application purpose in several real time disciplines.

References

1. Lian, K. L., Jhang, J. H., and Tian, I. S. (2014) A maximum power point tracking method based on perturb-and- observe combined with particle swarm optimization. *IEEE Journal of Photovoltaics*, **4**, 626-633,
2. Mohanraj, K., and Yokesh Kiran, B. (2019) PV integrated SEPIC converter using maximum power point tracking for AC loads. *International Journal of Recent Technology and Engineering*, **8**.
3. Danila Shirly, A. R., Vairamani, K. R., Vijayalakshmi, S., Dheepa, A., and Suganya (2016) Analysis of perturb & observe and incremental conductance MPPT techniques for solar powered SEPIC converter. *International Journal of Engineering Science and Computing*, **6**(3), 2906-2913,
4. Suganya, V., Danila Shirly, A. R., Vairamani, K. R., Vijayalakshmi, S., and Dheepa, A. (2016) Solar powered battery charger using sliding mode nontroller. *International Journal of Engineering Science and Computing*, **6**, 3012-3018.
5. Adhikari, N., Singh, B., and Vyas, A. L. (2011) Performance evaluation of a low power solar – PV energy system with SEPIC converter. *IEEE Ninth International Conference on Power Electronics and Drive Systems*.
6. Dheepa, A., Danila Shirly, A. R., Vairamani, K. R., Vijayalakshmi, S., and Suganya, V. (2016) Comparison of synchronous SEPIC and synchronous super lift LUO converter for photovoltaic energy generation system. *International Journal of Engineering Science and Computing*, **6**, 3031-3036.
7. Srinivasa Rao, T. C., Unnisa, F., Shobha Rani, D., and Srinivas, N. (2013) Photovoltaic power converter as an input source for SEPIC converter. *International Journal of Computer Applications*, 78(13).
8. Mahapatro, S. K. (2013) Maximum power point tracking of solar cell using buck-boost converter. *International Journal of Engineering Research & Technology*, **2**(5).
9. Dunia, J., and Mwinyiwiwa, B. M. M. (2013) Performance comparison between CUK and SEPIC converters for MPPT using

incremental conductance technique in solar power applications. *International Journal of Electrical, Computer, Energetic, Electronic and Communication Engineering*, **7**(12).

10. Sera, D., Mathe, L., Kerekes, T., Spataru, S. V., and Teodorescu, R. (2013) On the perturb and observe and incremental conductance MPPT methods for PV systems. *IEEE Journal of Photovoltaics*, **3**(3).

11. Banaei, M. R., Shirinabady, M. R., and Mirzaey, M. (2014) MPPT control of photovoltaics using SEPIC converter to reduce the input current ripples. *International Journal of Engineering Research and Applications*, **4**(1).

8

Power System Stability Enhancement using FACTS Controllers

R. Ramya, S. Usha, Depanjan Maiti, Siddharth Manna and Hrishikesh Sahu

Department of Electrical and Electronics Engineering, SRM Institute of Science and Technology, Kattankulathur, Tamilnadu, India

Abstract

In the past twenty years, the necessity of power has accumulated exponentially. The diversification of power production has been constricted because of limited resources and regulation of environment. Due to which, power lines are being burdened and therefore, the stability of system becomes a limiting factor for distribution of power. Flexible AC transmission systems controllers are used in such conditions. These devices have been typically used for resolving numerous issues of power system. This project is about the transient stability analysis and its improvement using a FACTS controller device like Static VAR Compensator, which is presented for two area and three area systems. In this project we have studied transient state stability with the help of MATLAB and improvement using FACTS devices like SVC using SIMULINK on these systems. In those systems due to any fault, the chances of losing synchronization and thus the risk of leading to a cascade effect on all machines are very high. The simulation results will reveal that use of FACTS devices like SVC is an effective solution to damp low frequency oscillation for such power system under the condition of fault.

8.1 Introduction

Successful performance of a power system depends largely on the engineer's ability to supply reliable and sustained service to the loads. The authenticity of the power supply implies far more than merely being available. Ideally, the loads must be continuously fed at constant voltage and frequency.

The first prerequisite of reliable service is to maintain the synchronous generators running in parallel and with requisite capacity to satisfy the load demand. Synchronous machines do not indisputably fall out of step under normal conditions. If a machine tends to pace up or decelerate, synchronizing forces tend to retain it in step. Conditions do arise, however, like a fault on the network, failure in a piece of appliance, sudden implementation of a crucial load such as loss of a line or generating unit, during which the performance is such that the synchronizing forces for one or more machines might not be adequate, and little impacts in the system may cause these machines to drop synchronism. A second prerequisite of reliable electrical service is to maintain the integrity of the power network. The high-voltage transmission system associates the generating stations and the load centers. Interventions in this network may hinder the flow of power to the load. This usually requires a study of large geographical areas since almost all power systems are interconnected with neighboring systems.

Random changes in load are happening at all times, with subsequent adjustments of generation. We may check out any of those as a change from one equilibrium state to a different state. Synchronism frequently could also be lost in that transition period, or growing oscillations may occur over a transmission line, eventually resulting in its tripping. These problems must be studied by the power system engineer and fall into the heading "power system". The problem is concerned with the comportment of the synchronous machines after an interference. Power system stability may be predominantly elucidated as the peculiarity of a power system that validates it to remain in a state of operating equilibrium under normal operating condition and to reclaim an admissible state of equilibrium after being subjected to an intervention.

8.1.1 Types of Power System Stability

- Steady State Stability - It is interpreted as the potentiality of the system to escort itself back to its stable configuration following a small disturbance in the network.
- Transient State Stability - It refers to the capability of the system to reach a stout condition following a large disturbance in the network condition. In all instances related to large changes in the system like sudden application or removal of load, switching

operations, line faults or loss due to excitation the transient stability of the system comes into play.

- Dynamic Stability - It denotes the artificial stability given to an inherently unreliable system by automatic controlled means. It is conventionally perturbed to small disturbances lasting for about 10 to 30 seconds.

8.1.2 Recent Problems in Power System

- Demand of power has been grown significantly with the elaboration of power production and transference has been restricted due to scarcity environmental resources and other limitations.
- Power lines are burdened due to which the stability of system becomes a limiting factor for transferring power.

The solutions to the above power system problems are as follows:

- Enhancing the capability of power transferring.
- Stability improvement transient and dynamic conditions.
- Improving flexibility in operation and control of the system

8.2 Description of FACTS Controllers

FACTS is an abbreviation of the "Flexible AC Transmission System". It is the utilization of electrical energy in a transmission system. It is an AC transmission system that consolidates an electronic power control and other stationary controls to take control of power transmission. FACTS control is defined as power-based electronics and other uniform devices that provide command of the framework of one or more AC transmission systems. In the early days of Circuit Breakers like Relay, Contactors were used to superintend the energy flow of transmission systems. Circuit Breakers were very unreliable and could not pay for power losses due to Reactive Power transfer systems. There are various types of FACTS controllers. The different types and classes of FACTS Controllers are categorized and shown in Figure 8.1.

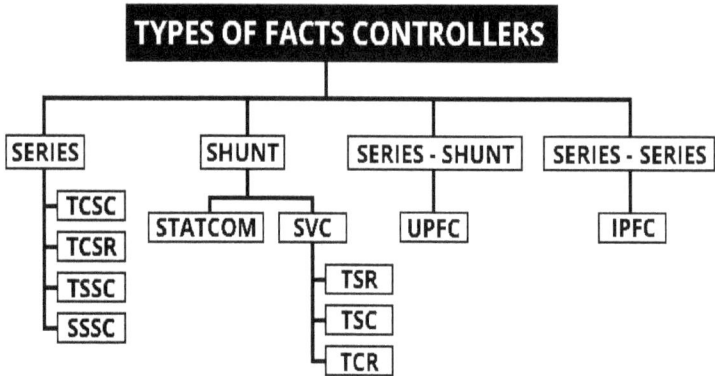

Figure 8.1 Classification of FACTS controllers.

8.3 Static VAR Compensator

Static VAR Compensation (SVC) represents collection of galvanizing instruments that provide instant active power in high-power dissemination networks. SVCs are proportion of the Flexible AC transmission device family, which controls electrical power, power factor, and harmonics and strengthens the system. The steady compliant compensator has null key mechanical hunks. Before the establishment of this expender, the compensation intermediary involved maintenance of enormous swapping units including concomitant condensers or reversed diodes to rages.

A single-line diagram of the standard SVC suspension is shown in Figure 8.2. In this illustration, a semiconductor diode-controlled coil-wounded unit, a crystal-rectified shifting capacitor, a consonant sifts, a mechanically replaced compensator, and instinctive modified reactor is used.

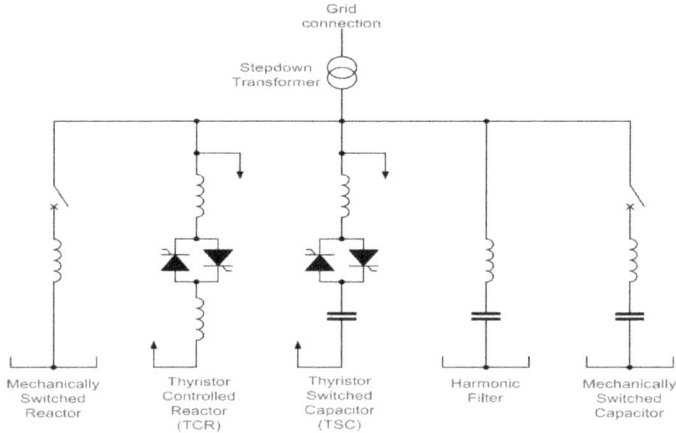

Figure 8.2 Schematic diagram of static VAR compensator (SVC).

SVC is mainly preferred for its great usage. It increases the ability of the power lines to transmit power. It improves the power factor of load, which reduces losses in line and improves the overall capability of the system. SVC performs compensation by differentiating a transmission line.

8.4 Analysis of Multi-Machine System

8.4.1 Simulation Model of Two-Area Three-Bus Power System

In Figure 8.3, the Simulation model of the project is given which depicts the working of the model. It depicts a one-thousand-megawatt hydraulic generation power plant that is fed to a load center through long five hundred kilo volts, seven-hundred-kilometer transmission line. The load center is equipped by a five-thousand-megawatt R-load. The load is connected to a one-thousand-megawatt plant and a local generation plant of five thousand megawatt. The system has been initiated such that the line carries a power of nine hundred fifty megawatts that is near to its surge impedance loading. So as to regain stability of system in post fault faults, the transmission lines are parallel compensated at its center by a two hundred MVAR SVC.

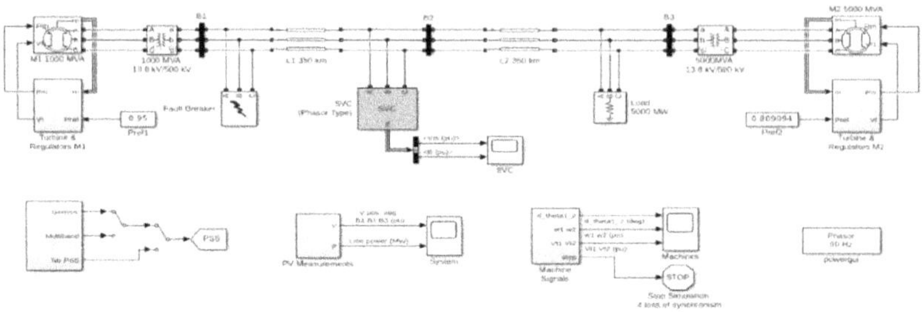

Figure 8.3 Simulation model of two-area three-bus power system.

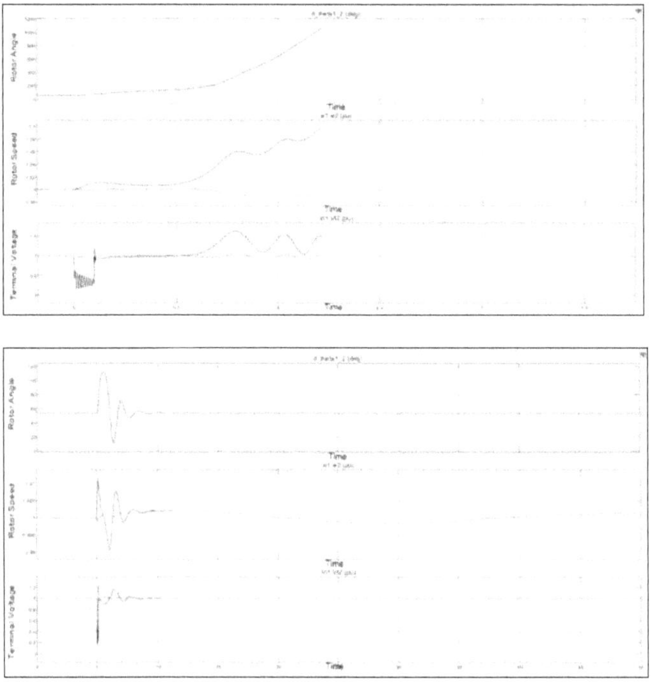

Figure 8.4 Machine performance (rotor angle, rotor speed, terminal voltage) of a system: (top) without SVC and (bottom) with SVC.

Figure 8.4(a) shows the different variations in the parameters without using SVC in the system. In the figure, the first graph depicts the variation in rotor angle which got deflected after 5 seconds, the second graph depicts the variation in speed of two machines which shows asynchronous behavior, and the third graph depicts the terminal voltage with respect to time during three phase faults with no SVC installed in the system. The simulation in this case has all the response till 6.25 seconds because of the stop simulation block used in the simulation system, which stops the simulation if there persists loss of synchronism for a certain rotor angle.

Figure 8.4(b) shows the graph of variations in the parameters with implementation of SVC in the system. In the figure, the first graph depicts the variation in rotor angle between 5 to 10 seconds and thereby a stable response, the second graph depicts the synchronism in speed of two machine with a little disturbance between 5 to 10 seconds, and the third graph depicts the terminal voltage with time which shows stability when SVC is installed in the system.

8.4.2 Simulation Model of Three-Area Nine-Bus Power System

In Figure 8.5, the Simulation model of the project is given which depicts the working of the model. A three-phase fault of occurs at bus B4 of 0.1second. The PSS is put into account by setting the value to 1. It depicts a one-thousand-megawatt turbine and generator that is fed to load center via a long five hundred kilo volts, seven-hundred-kilometer transmission line. The load-A and load-B are connected on both sides to the transformers through the transmission line. It has been initiated such that the line carries a power of nine hundred fifty megawatts that is near to its surge impedance loading. So as to regain stability of system in post fault condition, the transmission lines are parallel compensated at its center near load-C by a two hundred MVAR SVC.

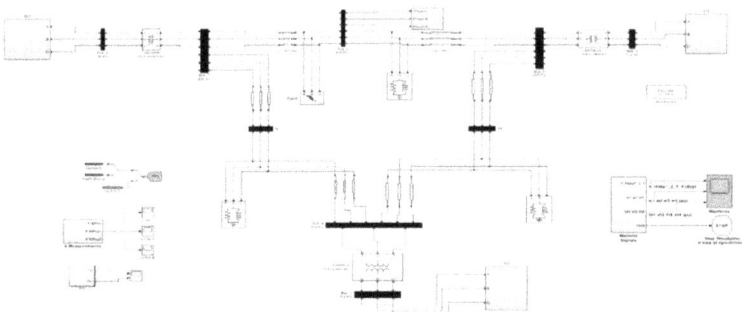

Figure 8.5 Simulation model of three-area nine-bus power system.

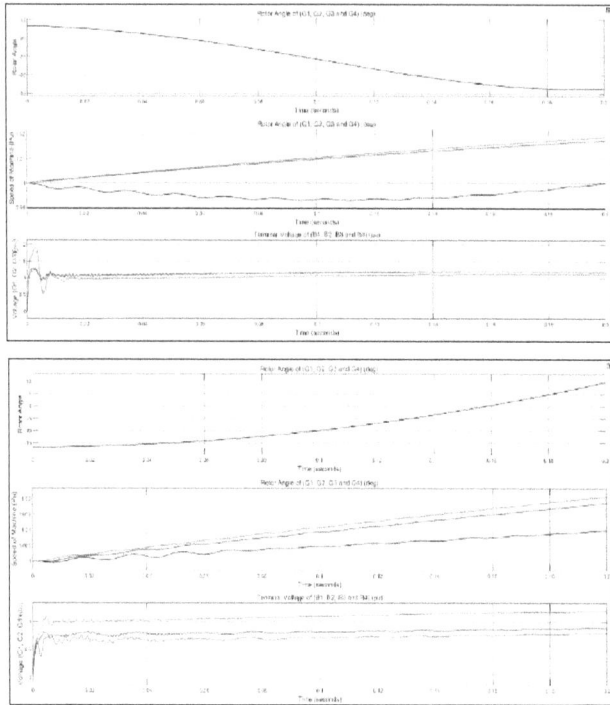

Figure 8.6 Rotor angle, speed and terminal voltage versus time: (top) without SVC and (bottom) with SVC.

Figure 8.6(a) shows the graph of variations in rotor speed, rotor angle and terminal voltage with respect to time. In the figure, the first graph depicts the exponential decline of the rotor angle of the system, the second graph depicts the variable machine speeds where the speed of one machine near the fault is affected and it lost synchronism, and the third graph depicts the distortion in the terminal voltage at starting with respect to time when SVC is not installed in the system.

Figure 8.6(b) shows the graph of variations in the rotor angle, rotor speed and terminal voltage with respect to time. In the figure, the first graph depicts the exponential increase of the rotor angle of the system, the second graph depicts the linear increase in the machine speeds with a slight disturbance in one machine which is near to fault, till 0.06 seconds and gaining stability thereby; and the third graph depicts the slight distortion in the terminal voltage at the beginning with respect to Time and how slowly it stabilizes after SVC being installed in the system.

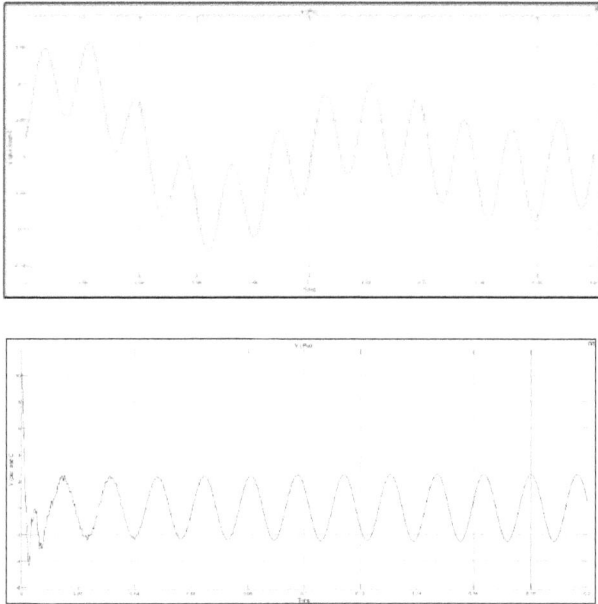

Figure 8.7 Response of load angle C vs. time: (top) without SVC and (bottom) with SVC.

Figure 8.7(a) shows the graph of variation in load angle occurring for load-C with respect to time when SVC is not installed in the system. The variation here is occurrence of non-linearity due to the fault being occurring near load-C as compared to other loads which are placed far from the fault position. Therefore, the graph of load angle C is not able to show stable sine wave nature.

Figure 8.7(b) shows the graph of variation in the load angle occurring for load-C with respect to time. Here the load–C achieves stability at less than 0.06s at approx. 0.048s after SVC being installed in the system. Here the load–C has been stabilized itself and it becomes linearly stable without much varying and thus, after implementing SVC, the stability of load–C is achieved.

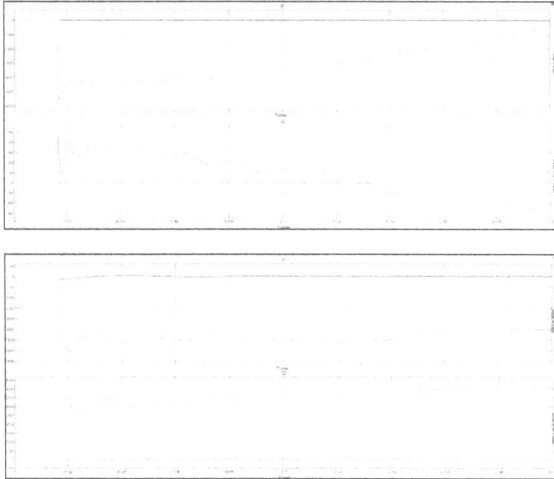

Figure 8.8 Real and reactive power vs. time: (top) without SVC and (bottom) with SVC.

Figure 8.8(a) depicts the graph of real and reactive power with respect to time, here it can be seen that due to the occurrence of fault the graph of real and reactive power intersects showing instability in the absence of SVC in the power system.

Figure 8.8(b) depicts the response of real and reactive power with respect to time, here it can be seen that the graph of real and reactive power does not intersect with each other in presence of SVC in the system which shows the stable condition in the power system.

8.5 Conclusion

From the simulation, we studied the transient stability and its effect over synchronism of a two area three bus power system and three area nine bus system effectively with and without SVC. We have mainly observed various parameters such as rotor angle, rotor speed, load angles, real power, reactive power and terminal voltage. In the two area three bus system simulation of without SVC, we observed a loss of synchronism when a 3-phase fault occurred. And in the simulation with SVC, we can see in the graph obtained that by the use of SVC, after a fault or disturbance, the power system has the ability to return to its normal conditions and gain the synchronism again. In the three bus nine bus system simulation of without SVC, three load angles were observed where the load angle near the fault has more disturbances and longer settling time for getting stable. And in simulation with SVC, we can observe from the graph that the settling time for stability of load angle has reduced effectively. So, we can conclude that the use of SVC is an effective methodology to improve voltage stability and to analyze the actual effect, the system is allowed to a 3-phase fault so that role of SVC can be evaluated in effective manner. We can conclude that using compensator we can stabilize our system up to a certain limit and maintain the synchronism and protect the whole system from cascade tripping.

References

1. Kaur. T., and Kakran, S. (2012) Transient stability improvement of long transmission line system by using SVC. *International Journal of Advanced Research in Electrical, Electronics and Instrumentation Engineering*, **1**(4).
2. Acero, J. F. C., Viamonte. W. R. L., Velasquez, O. C., and Pareja, W. O. P. (2020) Power quality improvement with SVC in a power system of 220V. *IEEE PES/IAS Power Africa*.
3. Joshi, N., and Nema, P. (2019) Voltage Stability Enhancement of wind based distributed generation system by SVC. *IEEE, ICSSIT*.
4. Hammad, A. E. (1986) Analysis of power system stability enhancement by static VAR compensators. *IEEE Transactions on Power System*, **1**(4).
5. Khan, S., Meena, R., and Bhowmick, S. (2015) Small signal stability improvement of a single machine infinite bus system using SVC. *IEEE Indicon*.
6. Kundur, P., Paserba, J., Andersson, G., and Hatziargyriou, N. D. (2004) Definition and classification of power system stability. *IEEE transactions on Power systems*.
7. Rai, J. N., Hasan, N., and Arora, B. B. (2014) Comparison of FACTS devices for two area power system stability enhancement using MATLAB modeling. *International Journal of Electrical Engineering and Technology*, **5**.
8. Bisen, P., and Shrivastava, A. (2013) Comparison between SVC and STATCOM FACTS devices for power system stability enhancement. *International Journal on Emerging Technologies*, **4**(2),101-109.
9. Jegedeesh Kumar, R., Arun, R., and Vignesh. T. (2018) Transient stability improvement of an IEEE 9 bus power system using FACTS devices. *International Journal of Innovative Science and Research Technology*, **3**(2).

10. Yome, A. S., Mithulananthan, N., and Lee, K. Y. (2013) Comprehensive comparison of FACTS devices for exclusive loadability enhancement. *IEEJ Transactions on Electrical and Electronic Engineering*.

11. Molazei, S., Farsangi, M. M., and Nezamabadi-pour, H. (2007) Enhancement of total transfer capability using SVC and TCSC. *Proceedings of the 6th WSEAS International Conference on Applications of Electrical Engineering*.

12. Mohanty, A. K., and Barik, A. K. (2011) Power system stability improvement using FACTS devices. *International Journal of Modern Engineering Research*, **1**(2).

13. Canizares, C. A., and Faur, Z. T. (1999) Analysis of SVC and TCSC controllers in voltage collapse. *IEEE Trans. Power Systems*, **14**(1).

14. Murali, D., Rajaram, M. and Reka, N. (2010) Comparison of FACTS devices for power system stability enhancement. *International Journal of Computer Applications*, **8**(4).

www.ingramcontent.com/pod-product-compliance
Lightning Source LLC
Chambersburg PA
CBHW050125240326
41458CB00122B/1406